Ocean Liner Sunset

Theodore W. Scull

Overview Press Limited, Mayes House,
Vansittart Estate, Windsor SL4 1SE UK
email: info@overviewpress.co.uk
www.overviewpress.co.uk

Contents

Acknowledgements

Most material comes from my personal journals written at the time of sailing, with additional research using shipping company brochures, deck plans and schedules and, of course, the Internet. I am indebted to my wife Suellyn and to Brenda Mayes for eagle-eyed proofreading; to Bill Mayes for once again taking on the tasks of cleaning up the photographs and publishing the book and Chris Mason for undertaking the layout. Thank you all.

ISBN 978 0 9547026 8 7 Published in February 2017. Printed by Amadeus Press, Cleckheaton, West Yorkshire.

Introduction

The third in the series on personal ocean liner travel recollections has to be the last and final volume as it is highly unlikely that the joys of point-to-point sea travel will experience a rebirth. Apart from a massive increase in leisure cruising, intercontinental sea travel, with just two exceptions, is an anachronism.

In 2017, the last true mail-carrying liner - *RMS St. Helena* - bows out following the introduction of a new air service that replaces the centuries old sea travel link between the remote British South Atlantic dependency of St. Helena and the outside world via the UK and South Africa. As a last hurrah, the ship received a temporary reprieve when excessive wind shear at the island's brand-new airport prevented the planned start of commercial air service on 21st May 2016. The other is Cunard's *Queen Mary 2* plying the Atlantic nine or ten months of the year inter-dispersed with cruises from New York, Southampton and Hamburg.

The ensuing chapters cover some of the most beloved liners on voyages as varied as P&O's *Canberra*, the largest liner ever built for a service other than the North Atlantic, and Polish Ocean Lines' ever-so-traditional *Stefan Batory* with a fond look back at no nonsense sea travel. We sail aboard the *Norway* on an eastbound transatlantic voyage, a complete break from her plodding Caribbean circuits, and I recall the same crossing when the ship was the French Line's *France*. Go aboard the SS *Rotterdam*, equally well suited as a two-class liner and a one class cruise ship on her most memorable final crossing, and P&O's superb *Oriana*, the last ship designed for the Orient Line, on her final liner voyage to Australia where she was to spend the rest of her seagoing days.

Passenger-cargo ships that combined the carriage of freight as well as a moderate number of passengers, and not just 12, were once all the rage. Here we sail on the Norwegian-flag *Americana*, an 86-passenger container ship on a dreamy northbound voyage from Buenos Aires via Brazilian ports to New York. No book devoted to passenger liner voyages could leave out the *Queen Elizabeth 2*, Cunard's longest serving ship, and one that was originally thought to be a misguided investment into a dying service. While her career was anything but routine, and she had her not so happy periods, the *QE2* will go down as one of the all-time greats. In this chapter, I recall her roughest crossing, and mine too. While I love a good blow on a ship that can handle it, the next storm does not have to exceed the one described in these pages.

Four chapters included in this third volume may seem a bit of stretch as the ships are not true ocean liners though they provide or provided regular departures from A to B, not between continents, but along lengthy coastlines: The Alaska Marine Highway's *Matanuska* from Washington State to Southeast Alaska; Hurtigruten's classic *Lofoten* connecting the south of Norway with the north beyond the Arctic Circle are two of them. A third service was Turkish Maritime Lines' venerable *Akdeniz*, withdrawn due to age, and a fourth, DFDS's failed attempt to operate the long-distance passenger and vehicle ferry *Scandinavia* as an alternative to driving between the north-eastern states and Central and South Florida. However, the requirement to change ships in the Bahamas because a foreign-flag carrier could not sail directly between two U.S. ports, killed the concept.

The last and final intercontinental sea link, unquestionably performed by a true ocean liner, is Cunard's *Queen Mary 2*, holding down what was once the most frequented water crossing on the globe, the transatlantic route between England, Continental Europe and New York.

She was constructed with all the design features needed to plough through any kind of weather with, except in the most horrendous sea conditions, the ability to maintain her transatlantic service speed. After a score of crossings in her, I reckon that she is the most stable liner ever built. At the sunset of a long era of ocean travel, she holds up her role in spades. Happily, for those who won't fly or for anyone who wants to sample a true ocean crossing, perhaps as their forbearers once did, it is still possible to cross the Atlantic by sea in a week.

Today, taking longer than a week getting to where you want to go would not be an option for most travellers. Britain to South African ports, once a major sea route, took a minimum of just less than two weeks to Cape Town and three to Durban; to Sydney, Australia from Britain, from three weeks to four or five; New York to Buenos Aires, roughly two weeks; and San Francisco to Hong Kong three weeks. All these journeys can be done in 12 to 20 hours by air. It's no contest anyway you look at it.

Enjoy an armchair Bon Voyage reading about a wide variety of the last liners in service between the years 1980 and 2016.

Ted Scull, New York, February 2017

Chapter One

Alaska Marine Highway's MATANUSKA

Wintry Northbound to Alaska

4th - 11th January 1980

Alongside Pier 48 on Seattle's waterfront, the blue and white ferry liner *Matanuska* was quietly resting between voyages. There was a light, cold drizzle falling over the city. Awaiting the signal to board, the foot passengers were huddled in one of two temporary waiting rooms, while the vehicle passengers sat snugly in their cars, campervans and pick-ups.

January seemed an odd time of the year to think about taking a seven-day, six-night voyage 2,460 miles up the Inside Passage to Skagway and back. "There won't be many tourists on this trip", said the warmly wrapped reservations clerk, who was adjusting the height of the boarding ramp.

The Alaska Marine Highway operates nine passenger ferries, all named after glaciers, over several different routes to complement the state's underdeveloped coastal road system. On the weekly Seattle service - increasing to two sailings a week from May to September - the 408-foot *Matanuska* calls at seven Panhandle ports. Five of them - Ketchikan, Wrangell, Petersburg, Juneau and Sitka - have no road access to the outside world. At the northern end of the route, the port of Haines connects to the Alaska Highway via a 150-mile gravel road through Canada, and Skagway has a newly constructed road and an old narrow gauge railway leading to Whitehorse in the Yukon Territory.

Ever since 1963, when the State of Alaska's Department of Transportation began operating its first ships, the service has developed into a new way home for many Alaskans, a cheap means to get to Southeast Alaska for newcomers, and a moderately-priced (about $50-60 a day including passage and a cabin) alternative to expensive cruise ships for visitors wanting to see some of the world's most beautiful and unspoiled maritime and mountain scenery.

Unlike luxury cruise ships, running only during the crowded summer season, the Alaskan ferries allow stopovers and a chance to meet travellers for whom Alaska is a destination, not a series of short port calls.

Three of the ferries were in winter layup and undergoing repairs on the far side of Pier 48. The flagship *Columbia* appeared particularly smart with the twinkling lights of the Big Dipper shining brightly against the dark-blue background of the ship's tall funnel. All Marine Highway ships have representations of the state flag on their stacks. Next to the *Columbia* lay the *Malaspina*, slightly smaller, and a sister to the *Matanuska*. Both "M" ships were lengthened by 56 feet in the 1970s, adding much-needed space to the car deck and substantially increasing passenger and cabin

The Matanuska is docked at Ketchikan on 9th January 1980.

capacity. The third laid-up vessel was the *Tustumena*, a seagoing boat that runs on the choppy, Southcentral route to the Kenai Peninsula and Kodiak Island, and less frequently out along the Aleutian Islands chain, a potentially tempestuous passage with legendary mountainous seas.

Once on board the *Matanuska*, a tour of the facilities created a most favourable impression. On the boat deck, the observation lounge with huge wraparound windows afforded views in three directions. Amidships, a dimly lit lounge bar was furnished with plush red chairs, and at the stern, a cheerful cafeteria had windows on three sides.

The sleeping accommodation (which must be paid for in addition to the fare) includes 112 two, three and four-berth inside and outside cabins with and without bathrooms. The size, arrangement, type and number vary with each ship. Cost is based on size and number of berths - not location. Since the accommodation is designed primarily for Alaskans (who make two-and three-night journeys), the cabins tend to be functional rather than well appointed. Mine was a relatively spacious double, arranged for single occupancy, with a shower and toilet attached. Passengers make their own beds or not, and if you take the roundtrip voyage you may request a change of linens.

The highest deck, Bridge Deck, had a reclining-seat lounge for those without cabins and deck space for the backpacking and camping set. According to the lounge attendant it is wall-to-wall bodies throughout the summer months. Aft, on the same deck, a heated solarium allows passengers to view the spectacular mountain scenery in comfort. At night, the hale and hearty sleep here.

Winter layup in Seattle - from the left: Aurora, Malaspina and Tustumena.

The 1974-built Columbia is southbound to Seattle.

Down on the car deck, the crewmembers were congratulating themselves on having boarded all the paying vehicles. The last few cars to make it were so closely packed together that the occupants had to climb across the bumpers with their suitcases to reach the stairway up to the passenger accommodation. They appeared not to mind the inconvenience and were happy to be aboard at all.

The deck hand told me that 19 'pass' cars had to be left behind. With that, the hydraulic doors slammed shut, and the vibrations under foot indicated that the *Matanuska* was away promptly at 8pm on Voyage 1172, north to Alaska.

By the next morning the dank, grey weather, so common in the Northwest, had given way to a crystal clear day with the temperature rising into the 30s (Fahrenheit). The ship was threading her way through the fast-running Seymour Narrows, at times a less-than-quarter-mile-wide passage between Vancouver Island and the British Columbia Coast. Precipitation during the night had left a blanket of snow on the mountaintops. On the lower slopes, the snowfall was lighter, until at the water's edge it appeared as a fresh dusting. The waterway was quite busy with tugs pulling loaded and unloaded container barges.

An announcement from the purser broke the silence, "May I have your attention please? For the next 40 miles and two and a half hours, the *Matanuska* will experience some rough sailing conditions as the ship traverses Queen Charlotte Sound. It will not be dangerous, but the captain advises all passengers and their children to remain seated during the crossing. I repeat"....

The *Matanuska* was not built to seagoing specifications because of its overhanging car deck, and she did dance a bit and took a few waves over the bow. The seawater froze on the metal decks, making a walk outside a slippery business and not recommended. By sundown, at 4:30pm, the ship was once again in the sheltered Inside Passage, and except for a 10-mile stretch called the Dixon Entrance, would remain protected from the Pacific Ocean swells for the rest of the northbound voyage.

At dinner, some of the passengers in the food service line were grumbling about the dining room going cafeteria-style, so I asked the white-smocked cashier about this.

Above: the heated solarium provides relatively warm and sheltered deck space aboard the Matanuska northbound to Alaska in January 1980.
Below: Observation Lounge with Alaskans returning home after holidays in the Lower 48.

It's 9:30am and the sun has not yet risen while approaching Juneau in January 1980.

"Tomorrow after Ketchikan", she explained, "there will be fewer people on board, and the new crew will probably offer table service for breakfast and dinner. Unfortunately, the ships went over to the cafeteria style service to cut labour costs, and now with reduced help and a different serving set-up, the ships cannot offer formal dining when there are more than 250 passengers on board". I looked forward to the new arrangement, paid for my salmon steak and took my tray to an empty window table.

Food is an extra charge and three meals a day averaged about $22. The menus offered about half a dozen main courses, including fresh seafood, at lunch and dinner.

Ketchikan, the first call since Seattle, came early in the afternoon of the second full day. The small city of 12,000 inhabitants stretched for several miles along the waterfront at the base of a formidable mountain chain. It felt good to be able to take a brisk walk to the boat basin where most of the large salmon fishing fleet was in for the winter. The centre of town was further along and had a fair collection of turn-of-the-century wooden buildings.

Of the 326 passengers on board, 152 disembarked here, to be replaced by 77 new faces. The atmosphere was now more relaxed and table service did come to the cafeteria as promised.

The Wrangell and Petersburg calls came during the night, and by the following morning, Monday, the *Matanuska* was approaching Juneau, the state capital. A richly colourful dawn began at 9am and continued with slowly fading hues for more than an hour. The dock, adjacent to the centre, allowed several hours to visit the state museum, take a stroll along the main street, visit the Red Dog Saloon (I had a cousin who once played the piano there), and climb the wooden staircases to the upper residential town for a long-range view of the surrounding mountains and waterways and directly down to the city's waterfront. Juneau is far and away the most isolated capital amongst the 50 states, walled in by surrounding mountains, with no road access to the outside world, and none planned.

Later in the day, Jim Driscoll, the *Matanuska's* chief purser, shared with me the state's attitude toward the Marine Highway. "The representatives in the interior of the state, whose constituents do not depend very heavily on the Marine Highway, would like to see the service go away as it costs millions to subsidise every year. It's not fair because they have their state-funded highways and down here we are without roads. While most consider the operation a necessary evil, and one that will never disappear, it should be run as cheaply as possible.

"The people in the southeast who use it regularly want a quality service and that means things like waiter service in the dining room. This winter you have half the fleet laid up in Seattle and our best ship, the *Columbia*, is broken down. The *Taku* is still running out of Prince Rupert when she should be in drydock being given a complete rebuilding job like this one had two years ago. This is a fine ship now, but I don't see how the *Taku* can make it through the summer when we need everything that floats in the timetables. There are plumbing leaks and peeling paint".

North of Juneau, the passage along the Lynn Canal, a natural channel, brought fine panoramas of the Mendenhall, Davidson, and Rainbow glaciers, all fed by huge ice fields up in the mountains.

Just before the Haines arrival late that same afternoon, Mike Crosby, the senior deputy purser, came on the intercom with a bulletin, "Canadian Customs advises all travellers that blizzard conditions now exist on the Haines Highway. Travellers should not to attempt to drive the road tonight. The border station is hereby closed until conditions improve and until further notice".

The passengers preparing to leave the ship here had already gathered in the forward lounge to form a convoy. They all firmly agreed to stick together until they had all passed through the worst sections.

Darkness had fallen by mid-afternoon when the almost empty *Matanuska* reached Skagway, the turnaround point. The winds coming down from the Coast Mountains made the walk along the deserted main street slippery and tough going. The false-fronted Gold Rush-era buildings were either boarded up for the winter or closed for the night. The substantial-looking Golden North Hotel, where I once spent a night in a lovely brass bed, stood shuttered and empty. It was not a night to spend more than a half hour on the icy streets of Skagway.

Skagway, in 1900, boasted nearly 30,000 inhabitants. The discovery of gold in the Klondike Region of the Yukon in 1898 brought tens of thousands

It's a bright sunny day in early January along the Inside Passage aboard the Matanuska.

of prospectors here by sea from Canada and the United States. A narrow gauge railway was built to carry the gold seekers and their supplies over the mountains to connect with the Yukon River steamers. In the winter, the White Pass and Yukon Route continued to be the only land route out of town, until the road was completed in 1978, for Skagway's 800 inhabitants. I took that train in summer 1971, a nearly all-day trip to Whitehorse, Yukon Territory's capital, positioned astride the Alaska Highway. After climbing the spectacular White Pass and winding between lakes, the ride was monotonously scrub pine and bush country.

In January, Skagway feels like a dead-end place, while in summertime, it takes on the carnival appearance of a human zoo when the luxury cruise ships disgorge hundreds of passengers daily, all milling about until it is time to get back on the ship.

During the night, the *Matanuska* touched at Juneau's Auke Bay, a landing 14 miles north of the city, before working her way south and west to Sitka just in from the Pacific Ocean. The fourth day's run through the twisting Peril Strait brought the ship within 100 feet of land on numerous occasions. For the entire morning, there was not one dwelling seen along the evergreen shores of Tongass National Forest, the nation's largest with 16 million woodland acres.

White-headed bald eagles, our national bird, soared overhead, and an amateur bird watcher pointed out the young eagles, brown in colour until they reach three or four years of age. Alaska has five times as many of these huge birds as the combined Lower 48.

A school of porpoises followed the ship, a deer came down to the shore for a look, and two humpback whales leapt completely out of the water at close range. Nobody got a picture, it all happened too quickly.

At Sitka, a local school bus took passengers into the town centre, which had grown a lot since I first visited in 1959. Then, there was no Alaska Marine Highway, and I flew from Juneau in an ex-World War II PBY Catalina seaplane, occupying a seat in the former gun blister.

The Russian Orthodox Cathedral, a product of 19th century Russian occupation, had been rebuilt after a disastrous fire gutted the structure in January 1966. During our visit it was closed, a pity, as once inside there is a fine collection of icons and religious paintings.

In a peaceful forest setting at the edge of town rests one of the largest and best-preserved collections of totems in the world, with a story attached to every one. They sit in solitary splendour amongst much taller evergreens.

Striking snow-covered Mt. Edgecumbe rises across the water to the west of Sitka, looking very much akin to Japan's Mt. Fuji, and the story that I heard back in 1959 still circulates 21 years later. The residents rose one morning to seeing smoke pouring upward from the cone. Rumours soon spread that the long dormant volcano might erupt, so some adventurous souls sailed over there to see what's what.

They discovered to their relief that some pranksters had piled up dozens of old automobile tyres in the crater, poured petrol over them and set them alight. The town folks were mighty relieved to learn the cause of all that smoke.

The Matanuska is sailing northbound in the Lynn Canal to Skagway.

With the homeward rush now over, the *Matanuska* was lightly loaded on her southbound run via Petersburg, Wrangell and Ketchikan to Seattle. The remnants of a blizzard that paralysed Seattle enshrouded the ship as she ploughed through the grey waters along the British Columbia coast. A lookout stood watch huddled against the driving snow at the bow, and the ship's foghorn sounded regularly at two-minute intervals.

Despite the adverse conditions, and the only really inclement weather since leaving the port of origin, the *Matanuska* arrived back at Pier 48 right on time on Friday at 6:30am. The vessel would spend the day resting her engines and preparing for another sailing that evening - Voyage 1173 North to Alaska.

I spent the night in Seattle and the next day boarded the 'Empire Builder' for the two-day journey to Chicago, stayed a night there and continued on the 'Cardinal' to visit relatives in Hamilton, Ohio. A couple of days later I rejoined the 'Cardinal' overnight for Washington, D.C., connecting to the 'Yankee Clipper' for New York.

The January Alaska Marine Highway story would appear in more than a dozen newspaper travel sections, and the Washington Post liked the full account so much that the travel editor, Morris Rosenberg, ran it over two Sundays, something he had never done before.

The original trio, *Malaspina, Matanuska* and *Taku*, completed in 1963, are still in service over a half-century later, as is the *Tustumena* of 1964 and the *Columbia* built in 1974.

Chapter Two

Polish Ocean Lines' STEFAN BATORY

The Final, Classic Transatlantic Liner

16th - 26th August 1981

Poland's transatlantic passenger service was little known in Europe and North America when compared to those of the major players, yet its sole passenger ship, the *Stefan Batory*, outlived them all with the exception of Cunard Line's *Queen Elizabeth 2* and *Queen Mary 2*, and she became the final liner on the St. Lawrence Route to and from Montreal.

Under the Treaty of Versailles (1919), Poland received a stretch of Baltic coast including the then tiny village of Gdynia inhabited by an insignificant population of about 500, mostly poor fishing families. Here, Poland developed a brand-new port to rival Danzig (Gdansk), and by 1930 when a new Polish shipping line was formed, the Polish Transatlantic Shipping Company, Gdynia's population had grown to a sizeable 42,000. Passenger and freight services operated via Copenhagen to New York.

By 1934, the line that had taken on the official name of Gdynia America Shipping Lines Ltd, ordered a pair of 14,300-ton, twin-funnelled motor ships from the Italian yard of Cantieri Riuniti del'Adriatico. The 20-knot *Pilsudski* came into service in 1935 and the *Batory* followed in 1936. The pair brought new standards and considerable popularity to the line.

Sadly, the *Pilsudski* did not last long as, when war broke out on 1st September 1939 with the Nazi invasion of Poland, she was converted to an Allied troop carrier and all too soon succumbed to a German mine laid off the Humber on 26th November 1939.

At the onset of World War II, the *Batory* carried Britain's gold reserves from Scotland to Canada and then received much acclaim, safely transporting hundreds of British children to Australia in advance of an expected German invasion of Britain. Following the *Batory's* withdrawal, a plaque commemorating this event was later transferred to her successor, *Stefan Batory*.

Following an illustrious - and lucky career - during WWII, the *Batory* returned to transatlantic service from Gydnia via Copenhagen and Southampton to New York where she thrived during the immediate post-war sea travel boom.

The Stefan Batory is tied up at Tilbury landing stage on 16th August 1981 prior to sailing to Montreal via Rotterdam.

On 1st January 1951, the nationalised company became Polish Ocean Lines, and with Poland now a Communist-block country, Cold War politics and the 'Red Scare' saw the *Batory* denied docking privileges in New York. She sailed eastbound for the last time on 3rd April 1951.

She was immediately transferred to a new route between Gdynia, Southampton, Karachi and Bombay. When the 1956 Suez Crisis closed the canal, she was given an extensive refit, and in August 1957 she began operating from Gdynia via Copenhagen and Southampton to the St. Lawrence River ports of Quebec and Montreal, with Halifax as the western terminus in winter.

Withdrawn at the end of the 1968 season after completing 111 round voyages, she continued as an accommodation ship at Gdansk before a final 1971 voyage to the scrapyard in Hong Kong. As a replacement, Polish Ocean Lines purchased Holland America Line's 15,024-ton *Maasdam*, in a transaction that proved to be an excellent decision.

When first completed in 1952 along with her sister, the 1951-built *Ryndam*, a publicity folder stated that the two ships "have been designed to do for the transatlantic traveller what Henry Ford did for the motorist. Both vessels are born out of a demand for comfortable passage to and from Europe by budget-minded people. Now for approximately $20 a day, it is possible to travel tourist with virtual run-of-the-ship privileges in these attractive new vessels".

Because the Transatlantic Steamship Conference required passenger vessels to carry at least two classes, unless the accommodations were designated all first class, Holland America complied by building the pair with only 39 first-class berths and 854 in tourist class. Three-quarters of the latter were two-berth cabins, a huge breakthrough for couples or two friends travelling together. With most other lower-class accommodation being four berths or more, that often meant that husbands and wives were separated. The Holland America pair was phenomenally successful and paid for themselves in five years.

The *Maasdam*, renamed *Stefan Batory*, sailed from Gdynia on 11th April 1969 to start a second and happily very successful career. I knew the ship from my time working for Holland America on Pier 40 in the mid-1960s. When she was in port, I liked to spend my lunch hour aboard and take a seat in the lovely panelled, forward-facing, first-class lounge or in the adjoining starboard bar that had been created from a forward section of the Boat Deck. I would imagine being one of a small club of three-dozen passengers who would thoroughly enjoy riding out a stormy February 10-day North Atlantic crossing at a cost of just over $200.

In the 1970s steamship rates climbed rapidly, and comparing *Stefan Batory's* fares with those on the *Queen Elizabeth 2,* the former was less than half the price and took twice as long to make the crossing. I calculated I would get more than four times the value for the same money. By then I had made five *QE2* crossings, and after a rousing start, I found her not worth the money until after she was re-engined and re-crewed in 1986-1987.

In March 1981, I booked the *Stefan Batory* for a mid-August westbound sailing and was offered one of the two remaining lower berths in an inside cabin on A-Deck for a passage from London Tilbury via Rotterdam to Montreal.

I spent that summer in Europe travelling and lecturing aboard P&O's *Oriana*, and then a few days before sailing from Tilbury, I stayed with long-time friends, Laurence and Jennifer Dunn, who lived just across the Thames in Gravesend.

The Gravesend-Tilbury Ferry ties up at the landing stage astern of the Stefan Batory on 16th August 1981.

Above left: this English family is migrating to a new life in Canada aboard the Stefan Batory. Right: before the weather warms deck chairs line the Promenade Deck. Below: passengers gather along the enclosed promenade. Opposite: activities on Stefan Batory's top deck include swinging within a giant life ring.

On 16th August, the three of us went over to the 'balcony', a local vantage point for ship watching, to see if my ship had come in. Indeed she had and she looked a fine specimen of a traditional liner, albeit rather small. A horizontal white stripe ran from bow to stern nicely dividing the black hull. The superstructure gleamed white, and the tall pale-yellow funnel had a broad red band, two narrow white bands, bold red letters 'P-O-L', and a red trident. She looked trim and inviting. Her profile had been altered; especially the bridge and forward superstructure, and she appeared more modern and streamlined. Short wings had been added to the funnel, and two less stylish masts had replaced the original fore and aft ones.

The embarkation notice issued by the London agents, Stelp and Leighton, stated that "Passengers must report at the Tilbury Landing Stage at 1330 hours prompt".

I boarded the little British Rail ferry *Catherine* on the Gravesend side at 1:15pm, and seven minutes later landed a few feet to the stern of the *Stefan Batory*. The landing stage is a floating dock designed to rise and fall with the considerable tidal variations. Long arms connect the facility to shore-based port buildings and British Rail's Riverside Station. Regular trains connected the landing stage with London's Fenchurch Street Station in a little over half an hour.

The huge embarkation hall was originally built to handle large numbers of passengers and their accompanying baggage for regular sailings of New Zealand Shipping, Orient Line, P&O and Royal Mail to the Far East, Australia, New Zealand and South America. Now there were only a few sailings a month by Soviet ships, British India's school ship *Uganda*, *Stefan Batory* and the occasional cruise ship. In Gdynia 330 passengers had boarded for this voyage, and Tilbury contributed a further 216.

In the cavernous hall, ticket inspection, immigration formalities and restaurant reservations passed quickly and within half an hour I was crossing the gangway. Entering the main deck foyer, the ship revealed her considerable charms with lots of shiny brass, fine wood panelling, oddly shaped light fixtures and thick plush carpets. One felt at home straightaway.

STEFAN BATORY
PLO

CHAPTER TWO

A steward escorted me to Cabin 328 where I met my roommate for the next ten days, an artist from Edmonton, Alberta. We set about divvying up the modest amount of closet and drawer space, and we agreed to dispatch all belongings not wanted on the voyage to the baggage room.

There was a single washbasin, a writing desk and stool. The sheets, pillowcases, blankets, and towels were all attractively marked with the Polish Ocean Lines emblem. A particularly nice touch was the diamond-shaped opening in the blanket cover showing the company logo on the woollen blanket beneath.

Ega, the cabin stewardess, took great pride in seeing that her block of cabins was kept clean and tidy. Her efforts made the tiny living space seem larger. The bath, shower and toilet facilities were along the corridor, but my cabin mate did not believe in showering. However, when he fell ill, I suggested that he would feel better if he soaked daily in a good hot bath. Happily, he took up the practice.

A selection of cabins aboard the Stefan Batory taken from contemporary brochures.

At promptly 4pm, with the aid of two Alexandra Towing Company tugs, the *Stefan Batory* began to edge away from the landing stage. The simple manoeuvre involved a 180-degree swing to point the bow downstream, and the strong outgoing tide called for the tugs to use their full horsepower. On the far shore, I could just make out Laurence and Jennifer waving from the 'balcony'.

Once turned, for the next hour and a half, the *Stefan Batory* headed down the ever-broadening Thames between the low green hills of Kent and Essex. We passed a small inbound container ship named *Lusitania* and several tankers discharging crude oil. Then at Southend, we passed the more than mile-long pier, which jutted out into the deep water and built to receive excursion vessels. In the mouth of the estuary, rusting antisubmarine platforms were reminders of a war now nearly four decades in the past.

At 5:30pm the river pilot left via a rope ladder to the awaiting Trinity House launch. Now well into the open waters of the southern North Sea, the ship increased her speed to 16 knots for the overnight crossing to Rotterdam.

A VW Westfalia camper is hoisted aboard during the call at Rotterdam.

Dinner that evening was a delightful first experience that gave every indication of making mealtimes focal points of the day. The restaurant was largely unchanged from the days when the ship was under the Dutch flag. In the shape of a horseshoe, it could handle up to 400 passengers at round and rectangular tables seating from two to ten. In the evening, a four-piece band played classical and popular music against a backdrop of inland wood panels.

The people assigned to my rectangular late seating table were a good representation of the English-speaking passengers aboard. Sue Howson, from Toronto, was returning from two years working with the Bank of England. She had grown up at Steamer Point in Aden, and as a little girl, used to delight in visiting the Anchor, Orient, P&O and Bibby liners that called there en route to India, the Far East and Australia. Her husband, Donald Moggridge, had completed two years of teaching and research at Cambridge. Dorothy Wigmore, from Sudbury, Ontario, was taking home an English university degree in factory safety. Virginia Clark had been a librarian in London and was returning to Connecticut to continue editing a magazine for university professors. Mel Strong, from New Zealand, was making a tour of the world and had his sporty Triumph motorbike in the hold. Graham and Genia Westgate, from England, were emigrating to Canada to take up a job with Pratt and Whitney in Montreal.

While my table represented people who liked ships, they also benefited from the reasonable fares and considerable free baggage allowance when moving house and home between continents.

The daily menus presented a good choice of Polish and Continental European cuisine at both lunch and dinner. For starters, a hot cream borsch, Portuguese sardines, and a mixed vegetable salad were favourites; potato blinis with cream were popular, too. For the fish course, I liked the red snapper and the more exotic boiled eel in dill sauce. Main dishes included roast duck, veal in a white wine sauce, ragout of lamb, and medallions of beef tenderloin with mushrooms. The cheeses tended to be on the mild side, and simple desserts ran to a sour cherry mousse, gingerbread in chocolate sauce or a variety of ice creams. The chefs did an admirable job of varying the selections and making meals most pleasant occasions.

The wine list had reds, whites, and rosés from Bulgaria, Tunisia, Germany, Spain, Italy, and France beginning at $4.00 a bottle for a drinkable Bulgarian cabernet. All transactions were in U.S. Dollars, even though the ship never called at an American port. The

Above: the orchestra played at both seatings of dinner as well as tea time and in the evening.
Below: our table of ten, with two missing including the author/photographer, represented New Zealand, Canada, Britain and the U.S.

The author aboard the Stefan Batory.

Stefan Batory was a foreign currency earner, which, given Poland's desperate financial situation at that time, was probably providing no more than a drop in the bucket of crippling debt.

Ryszard Pytlik, the Head Chief Steward, who traced his sea career back to the days of the old *Batory*, described the food as good, but certainly not as exciting or varied as when the line had a first class. The *Stefan Batory* received top priority by the Polish government, and if there was a temporary shortage, stores could be replenished in Tilbury, Rotterdam and Montreal.

At Rotterdam, the ship tied up all day at Wilhelminakade, the former Holland America Line pier, crane-loading steamer trunks, large wooden crates, tea chests, automobiles and camper vans.

After watching some of the dockside activity, I went by subway to Rotterdam Central Station, boarding an inter-city train for a day trip to The Hague and finally tram to Madurodam, an outdoor city and country model built to a scale of 1 to 25 bustling with streets, canals, buildings, ports, trains, planes and ships. During my last visit in 1958, the featured ship was Rotterdam Lloyd's *Willem Ruys* (later *Achille Lauro*), and now it was Holland America's splendid *Rotterdam.*

The *Stefan Batory* departed Rotterdam promptly at 4pm, after embarking a further 228 passengers. For this westbound August sailing, the ship would be just four short of the 779-berth capacity. That was down from 881 when she was the *Maasdam,* with 102 berths eliminated during an extensive refit in 1968-1969. Because of the male/female mix, several young women ended up bunking down on mattresses in the former first-class restaurant up on the Boat Deck.

Apart from this heavy sailing, bookings had suffered because of Poland's political situation. Fewer overseas Poles were travelling to Poland for family visits, and not many native Poles were making the round trip to Canada and back. The increased number of passengers from England and the Continent made up some of the difference.

During the two-hour, 25-mile transit of the New Waterway, we had a ringside view of the world's busiest port. Both banks were lined with dozens of traditional break-bulk cargo and newer container vessels, huge

The dogs have lots of space to be playful on the Stefan Batory's top deck and passengers are enjoying the fine North Atlantic weather on the Boat Deck. Opposite: deck chairs grouped on the Boat Deck aft, and the author is practically galloping through his constitutional on the Boat Deck.

tankers and extensive ship repair yards. We met and overtook hundreds of powered barges hauling goods into and out of the heart of Europe via the Rhine.

At the western end of the New Waterway, the Sealink ferry *St. George* was berthed alongside the Hook of Holland railway station, awaiting the arrival of express trains from all over Europe. She would then make a nine-hour overnight crossing to Harwich where the sea service would connect with the "Hook Continental" boat train for London's Liverpool Street Station. Passing into the North Sea, the Sealink ferry *St. Edmund* was arriving after a daytime dash across from Harwich.

All day on 18th August, we steamed in bright sunshine through the English Channel along the South Coast of England passing close to the Isle of Wight, Portland Bill, Bolt Head, The Lizard and finally Land's End and the Isles of Scilly. The little *RMS Scillonian III* passed astern en route from St Mary's to Penzance. Then, with the Longships Lighthouse receding into the mist, the liner and her passengers were at last free of the land.

My own day invariably began with a wake-up call announcing the first breakfast sitting. Following that I began a two-mile constitutional on the boat deck, usually shared with other serious walkers; one, Captain Boleslaw Rakowski, who measured six-feet, four inches, easily overtook me with his long stride.

On different days, the sea assumed various moods ranging from fine sunny weather and crisp cool breezes to severe gales and heavy rain. On most days, the Atlantic simply appeared grey and overcast with daytime temperatures hovering in the low 60s.

A few brave souls managed a quiet dip in the tiled swimming pool aft on the promenade deck. On better days, the boat and promenade decks were crowded with passengers wrapped up in heavy blankets in true North Atlantic style, reading, talking or dozing. On poor days, most moved to their relocated deck chairs along the glass enclosed promenades sharing the space with older Polish men seated on folding wooden chairs discussing politics. The deck steward discouraged passengers from erecting their chairs by revealing two missing fingers to prove his point.

The liveliest area was the sports deck. Here, in all kinds of weather, owners of some 25 canine passengers frolicked with their pets, and after a few disorienting days, most seemed to enjoy the crossing.

The hours on deck passed blissfully with storm petrels the only signs of life, winging about, hundreds of miles from land.

In the morning, a recorded concert took place in the Grand Lounge, and a notice set in the middle of the dance floor, asked for silence. The passengers were a disciplined lot and a good crowd gathered here every day to have morning coffee and listen quietly to the music. At teatime, a live band played and a bingo session followed.

The Grand Lounge had been the *Maasdam's* Palm Court, furnished with upholstered bamboo chairs and decorated with long rows of tropical plant stands. Now the room had modern furnishings, and the port side had been partitioned off to create a coffee shop with booth-style seating.

On the port side aft of the lounge, the cherry-wood panelled library retained its décor and layout, with small selections of books in Polish, English and French and daily news sheets.

Some cabins had been removed to create a Chess Room on the port side and a Card Room on the starboard side. On the Promenade Deck, in place of the former Card Room, were two attractive additions, the

Polish Inn, a light-wood panelled bar with stools and the Hunters Room, a darker room with individual seats surrounding a large round table and the facsimile of a central fireplace.

The Little Lounge, panelled in Bubinga wood and used for dancing at night, once doubled as the tourist class theatre. Since then, a permanent cinema had been constructed in the former forward hold.

Totally unchanged was the dark, cosy walnut panelled American Bar, a semi-circular room with bar stools and several tables. Holland America was well known for its intimate windowless bars. The Smoking Room's aft-facing windows were no more and the empty box-like room during the day came alive at night when dancing took over.

The entire deck above, except for the Children's Playroom (once for both classes), was originally penthouse accommodations for 39 first-class passengers in 15 twin-bedded outside rooms, with extra Pullman upper berths, and all fitted with en-suite shower facilities. The former dining room, closed since the ship went to one class in 1976, was now little used. The former playroom had been turned over to teenagers.

Forward was my number one favourite pair of rooms on any ship in that period. They had served as the first-class lounge and bar and were without parallel for their cosy genteel atmosphere. Regular passengers considered them to be a private club, and one was expected to conform to the unwritten rules of quiet during the day, lively conversation over drinks in the evening before dinner, and silence while listening to the piano, bass fiddle and violin trio playing medleys every night for an hour from 9:30pm.

The décor, of Italian walnut, Zeeland elm and inlaid wood designs, was all original. Groups of individual chairs and long couches were positioned to allow views of the sea. Mounted on the bulkhead to the right of two writing tables was a wonderful wall clock with ivory figurines in the signs of the zodiac on the face and protected behind a façade of clear plastic.

The tiny starboard side bar can seat fourteen in three groupings with five stools at the bar. The view is forward, to the side and through a large window looking aft down the boat deck. This bar was an addition in the very early 1960s and now overseen by a masterly steward who remembered what everyone liked to drink after the second day out.

Late afternoons were invariably spent in a friendly game of Scrabble. However, most of the time seemed to be spent determining whether I was allowed to use my American spelling, Dorothy her Canadian slang, Don his Latin terms at Cambridge, and Sue the colloquial expressions she had acquired in New Zealand. The 'Official Scrabble Dictionary' was voted down three to one because it contained only one brand of English. American English still seems to be considered a powerful threat to the rest of the English-speaking world.

The captain's cocktail party was conducted in three separate sessions in a manner unlike any other I have attended. After passing through the receiving line, passengers were escorted to a table and seated with others speaking the same language. Drinks, an odd-tasting punch of some sort, were served at your seat while the officers moved around the room spending exactly five minutes at any one table. Promptly at 6pm, the band struck up a marching song, the officers rose, walked to the door, turned around and bowed to the passengers and then left. The party was considered at an end.

The highlight came on Polish Night when talented members of the crew and a couple of professional artists in full costume presented an evening of folk dancing and singing in the Grand Lounge. A couple of nights later, the traditional fancy dress ball was won hands down by an amazing long slithering caterpillar filled out by three generations of a large Chinese family.

During late night walks on the boat deck, I witnessed some spectacular displays of the Northern Lights. The pattern of vertical reflections was constantly changing colours and shapes, and the brightness rivalled the light of the full moon. On one absolutely clear night, while gazing up at the mast and funnel pointing skyward, I had the sense of moving out into the universe standing on the outside of a slow moving space ship.

For a ship nearing thirty years of age, the *Stefan Batory* was kept in impeccable condition by a crew that was obviously proud to serve in her. The brass was polished, the glass cleaned, the carpets freed of stains the moment they appeared and the light fixtures dusted, all on a daily basis. The captain's morning inspection is by no means cursory, and the crew takes the review with the utmost seriousness.

Probably the most important overseer is King Stefan Batory himself, seen in a life-size portrait by all who pass through the main foyer. The dignity of his stance and the simplicity of his red, white and black garments are an example to everyone on board. Stefan

Batory was a popularly elected 16th century King of Poland and Transylvania, whose family had Hungarian origins and who was responsible for bringing the Jesuits to Poland. When the MS *Batory* (Batory is the family name) was retired from service, it was a logical step to name the newly-acquired ship after this specific king to maintain the continuity that this line is noted for.

I was invited to Captain's Rakowski's table for the farewell dinner. We gathered in the Polish Inn for drinks, and at the dinner bell we remained seated until instructed to rise and follow the captain in a kind of parade. We entered the restaurant, and with all the other passengers seated, the band struck up a marching song as we filed through to our table. It was all quite triumphant and great fun. Happily, most of our table were also invited.

On Sunday, 23rd August, after five full days at sea, we sighted the barren rocky promontory of Belle Isle just off the southern tip of Labrador. The desolate headland could have hardly looked like the Promised Land to many of the immigrant passengers seeing the New World for the first time. More upbeat, it was here that everyone got excited at seeing small ice cakes floating by at close range.

Interestingly, once in the huge Gulf of St. Lawrence, the coastline disappeared again, and for the next day it was like being at sea. The French-Canadians refer to the St. Lawrence as 'la mer' (the sea) right up to the mouth of the Saguenay River. Here we spotted some spouting whales and the St. Lawrence narrowed further. Numerous outbound ships passed close by, either domestic vessels trading to Newfoundland and Labrador or container ships and bulk carriers going overseas. A couple of ferries connecting the north and south shores crossed our course at right angles.

Small villages began to appear and cars and trucks could be seen travelling along parallel highways. By late afternoon on the 25th, the *Stefan Batory* passed beneath the imposing citadel at Quebec City with the majestic Chateau Frontenac taking centre stage amongst the cluster of traditional and modern buildings.

The south bank offered a less impressive view of the drab city of Levis; Via Rail's day train, the St. Laurent, paused here briefly on its way to Montreal. Passengers connecting to Quebec City cross the river by ferry.

The river narrowed still further, and when the *Stefan Batory's* forward mast seemed to barely clear the road and rail bridges just upstream from Quebec City, several hundred passengers gathered on the sun deck broke out with a spontaneous round of applause. Small boats came out of little harbours to pace the ship and local inhabitants came down the river to watch us pass.

There rose the distinct feeling that the journey was approaching its natural end. We were slowly making our way into the heart of a great continent and after a week at sea, my thoughts were reluctantly coming back to earth. There were bags to pack and passports, entry forms and onward rail tickets to check over again.

The last dinner became a joyful celebration for 29 meals taken together. We splurged by eschewing the rough red Bulgarian Zagore and moved up to a smooth French Beaujolais. After dinner we all stayed up much too late to watch the lights of Trois Rivieres pass to starboard. Montreal was now only hours away, and so were the long lines for immigration, the interminable wait while mountains of baggage were off-loaded, and the struggle to attract the attention of a customs inspector.

The end of a transatlantic crossing can be a rude shock, and yet perhaps it is the best preparation for a return to normal life. Later the same day, I boarded the Montrealer occupying a roomette bound overnight for New York.

The *Stefan Batory* would remain in transatlantic service until 1988 when expensive work to meet Canadian waste discharge regulations could not be justified. She later served in stationary roles as an accommodation ship and was finally scrapped in Turkey in 2000. A significant era in transatlantic sea travel quietly passed into history leaving a trail of happy memories.

The Stefan Batory is resting between voyages at Montreal on 26th August 1981.

Chapter Three

P&Os ORIANA

Last Line Voyage to Australia

18th November - 1st December 1981

On Tuesday, 17th November, I took a cab to the East Side Air Terminal, boarded a bus to JFK Airport, flew to Bermuda and hopped in a cab for the Inverness Hotel, Hamilton, located near Front Street. I checked in with the shipping agents and found the *Oriana* to be on schedule for the first leg of her last line voyage. It had been 21 years since the *Oriana* began sailing between the UK and Australia via Suez, the Cape and Panama.

The *Oriana* was the fastest ship ever to sail for P&O or the Orient Line, with a service speed of 27.5 knots. She was delivered at about the time that the two lines fully merged as the short-lived P&O-Orient Lines, and was second in size within that fleet to the slightly slower *Canberra*. This pair cut a full week off the Southampton-Sydney passage held down by the previous generation - *Arcadia*, *Orsova* and others - built in the 1950s. Now in *Oriana's* semi-retirement, she would be a full-time cruise ship sailing out of Australia.

The following morning I walked to the landing and watched the ship's first load of passengers who had boarded at Southampton coming ashore aboard the tender *Canima* (formerly the Cobh, Ireland transatlantic liner tender *Blarna* built in 1961 with the capacity to carry up to 1,000 passengers).

About 35 passengers boarded here, setting off at 11:15am for the ship anchored in Grassy Bay and happily much closer in than *Canberra* a few years earlier when the tender trip took 45 minutes.

I had cabin A-42 to Port Everglades, then C-31 to San Francisco. The first cabin, located just above the lifeboats, had a single bed located well inside the ship with a narrow corridor running down to the porthole, an arrangement known as Bibby-style. That design, in the days before air-conditioning, originated with the Bibby Line of Liverpool and permitted many more cabins to have ocean air access.

The ship sailed a bit late at 3:45pm, just behind the *Rotterdam*.

Roger Lee, the Cruise Director, had not received my lecture list but was happy to see me, as there were very few entertainers booked for the voyage. I was slotted in at 11am so as not to conflict with the art class. There was a bit of competition between lecturers as to garnering audiences, and when I arrived on the scene in 1979 with

The Oriana is anchored for the day in Acapulco Bay on 27th November 1981.

Top: looking forward from the Writing Room along what was the Oriana's First Class public room deck and this bird may be a bit startled about what he sees behind him. Writing Room SS Oriana.

my ocean liner talks, they drew people away from the art, crafts, and bridge classes and the port lecturer's talks, so to keep the peace, the cruise director tried to schedule us not to conflict with one another. It would take me a few years to be accepted by my veteran colleagues, after which all was well, mostly.

During a ship inspection, I found that the Midships Bar had been redecorated for the Australian trade with the addition of a brick floor and large beer kegs with Australian news clippings embedded under glass as tables. The ship, as built, already had a fine drawings collection depicting the strange endemic birds and animals that populated Australia and New Zealand.

On the first day at sea, I helped the assistant cruise director with scoring the Syndicate Quiz, where tables of six competed with one another. The almost daily sessions were good fun and would become very competitive on a long voyage such as this one. I was allowed to participate, and later the table I joined won several times. My American knowledge came in handy on this boatload of mostly British, Australian and New Zealand passengers.

The call at Port Everglades saw scores of Brits disembark with Aussies and Yanks replacing them. Upon leaving, one local homeowner well known to the officers, flashed his lights as a farewell as did owners of a couple of dozen apartments. Additionally, several horns from ashore got into the act, plus a stirring rendition of *Rule Britannia.* Our own booming steam whistle reverberated off the buildings, then all fell silent as we passed beyond the channel jetties and out to sea, a last and final call for this veteran liner.

My new cabin on C Deck was referred to as a court cabin. The single was in a widening corridor of a half-dozen others, with a vertical slit window that looked into a court that ran to the ship's side and a large oval window. That way one had some natural light and, in the morning, one could eye the state of the sea. The space at the end of the court became an informal gathering place, furnished with a padded bench and low table, often used to display the overflow of bon voyage flower arrangements. My second cabin was not as nice as the outside single but perfectly adequate and with en suite facilities.

The following morning we docked at Nassau, sharing the port with the *Dolphin*, *Emerald Seas*, *Amerikanis*, and *Sunward II.* I boarded *Amerikanis* to seek out the remaining bits that reflected its first life as *Kenya Castle*; mostly the layout but not many decorative features. I had already visited the *Emerald Seas* and *Sunward II* and had little interest in the *Dolphin.*

I then ambled over to the nearby Sheraton British Colonial where my parents had stayed in the late 1930s, and found the hotel's exterior better reflecting its pre-war past than the modernised interior. In the less developed former steamer days, the hotel had an enviably convenient location but now it was a bit too close to the hoi polloi.

My forward cabin, placing it in former first class section, was allocated to the original first-class Elizabethan restaurant where virtually every officer down to the third rank, not on duty or asleep, hosted a table at dinner whilst the ship was at sea. I sat with

the Deputy Catering Purser, two New Zealanders and an Australian, and we had a jolly time of it. The three paying passengers were returning home at the tail end of their long-service leaves, usually granted after ten years on the job, and lasting from two to four months, depending on the Aussie state.

On Sundays at sea, British seagoing tradition had the captain, in this case Deputy Captain Mike Moulin, officiating at what is called a non-denominational worship service, but in fact is straight out of the Anglican prayer book, and very similar to the Episcopal service I grew up with. On long liner voyages, a passenger-organised choir would give the service a big boost. I did a little prayer for my first lecture.

I had a near full house for 11:15am, with the topic Transatlantic Liners Part I. Everything went well except that the vertical photos did not fit the screen. Numerous people came up afterwards to have a chat and share stories, such as having been one of the thousands of troops aboard the *Queen Mary* or *Elizabeth* during World War II. Another quite elderly man had been one of eight gardeners aboard the *Berengaria*, a former Cunard liner.

The air was now hot and humid on approach to Cristobal, passing amongst many anchored ships inside the breakwater, to reach the pier for docking at 7pm. After dinner I went ashore with a couple of others and came upon a group of shipping offices frozen in time; some still marked Italian Line, KNSM, United States Lines, Royal Mail and Pacific Steam Navigation, none of which still operated.

As a precaution drilled into me by a friend who worked on ships, I had my driver's license tucked in my front shirt pocket for identification and 11 dollars in my jeans' back pocket, with everything else left aboard. Two guys and a girl passed us and a few minutes later, one of the men suddenly came up to me and ripped open my shirt pocket and fled with my license. When he eyed what he had snatched, he crumpled it up and threw it on the ground. We had gone too far from the centre, so we doubled back passing a couple of policemen but did not say a word. Why bother?

On 24th November, I was up at 7am, invited by Captain Malcolm Rushon and Deputy Captain Mike Moulin to come up to the bridge for the canal transit. I was thrilled. *Oriana* took her place as number eight in a convoy of twenty ships. At first, the long approach had thick rain forests either side. Remnants of the French Canal opened up to the right, and the beginning of a parallel canal started during World War II ran off to the left. *Oriana* boarded two American pilots, and P&O signed over complete financial responsibility to the

Opposite, top to bottom: an enclosed promenade aboard P&O's Oriana lost its popularity as a windy weather refuge when the warm days arrived. Another section of the enclosed promenade portside aft. Princess Alexandra christened the Oriana at the Vickers-Armstrong yard on 3rd November 1959. Above: passengers are embarking in the late afternoon after a day ashore in Fort Lauderdale.

Canal Company. I believe that this is the only pilotage that demands and assumes all risk.

One of the two pilots told me that a heavily loaded container vessel with an aft superstructure carried four pilots to be able to see the ship's exact position when not visible from the bridge.

The Gatun Locks are made up of three chambers, each 1,000 feet long and 110 wide, and the minimum depth is 40 feet. For a ship of *Oriana's* size, six electric mules guided us through, operating in relays. Gatun Lake, into which we sailed, is fed by the Chagres River for a gravity flow operation. A spillway allows excess water to flow into the Atlantic. Locks fill in 10 minutes, yet it took about 2 hours 40 minutes to pass through three chambers because of the heavy traffic and slow pace of advancing the ships.

Oriana's bridge wings once knocked six tiles off one of the control tower roofs, and now these roofs are fitted with a hinged mechanism to be able to retract a few feet. *Oriana's* transit fee had been $55,578.50 in April 1981 while UK-bound, *QE2* paid $89,154, the record amount, in 1980. Author/adventurer Richard Halliburton paid just 36 cents when he swam the canal's length in 1928.

The speed limit is nine knots while passing amongst the lake's islands, and we reached Gamboa, division headquarters at 1:30pm. Here we began running in parallel with the Panama Railroad, built in 1855 and once the busiest in the world in traffic carried per mile. The Gaillard Cut follows, a narrow single-file passage between high ground either side, with Gold Hill to starboard.

Heavy showers drove us inside the bridge for about 30 minutes after which the skies remained overcast. In the Pedro Miguel Lock, an Ecuadoran freighter *Isla Baltra* had broken down and had become top heavy having used up its fuel. As it was currently in an unstable condition, tugs towed it out of the way. Then Miraflores and the final two chambers, after which *Oriana* passed under the Thatcher Bridge, which carries the Pan American Highway that begins in Alaska and is interrupted by impassable geography at the Panama/Colombia border. The bright lights of Balboa and Panama City were seen off to port, and finally, with the pilots dropped, we headed out into the Pacific - free at last. The officers retook the command of their ship; the air was fresh and the sea calm.

Above: the Oriana, soon to follow, is docked at Port Everglades while Chandris' Amerikanis begins her overnight sailing to Nassau.
Below: on another voyage Captain Dennis Scott-Masson on the bridge wing.

Part II of my lecture on transatlantic liners took place in the Plymouth Room, and the packed house made it impossible to provide everyone clear sight lines, even with the remote control cord at its maximum length. The screen mount was on a newly expanded stage that the crew dubbed Weymouth Pier. During the talk, I got lots of laughs and felt at my best. The third lecture, Beyond the Atlantic, covered British Colonial and Commonwealth liners and also drew a great response.

On 27th November, we dropped anchor in Acapulco Bay, but with one gangway out of action, a horrendous back up developed, and I did not disembark before 11am. Once ashore, I did not take to the place at all, too much of a shopping market, so I returned to the launch and snapped some splendid photos of the ship.

During the afternoon, Cyril Smith, the coxswain, had a heart attack and died at the wheel of his launch. He had been with *Oriana* since she came out in 1960,

Amerikanis, Oriana and Sunward II at Nassau.

and he was well known to the regular passengers. He featured in the promotional film that introduced the ship to the general public. Sadly, he had planned to retire after this last voyage and had recently bought a house in New Guinea.

Deck tennis took up a lot of time during the days at sea with a regular group meeting in the morning and the afternoon, occasionally including an off-duty officer. These sessions were a lot of fun and a good way to meet others and hear their stories. One, an English woman, had been on many ships as a passenger, but she had always wanted to work on one. When she was starting out only women aged over 45 were hired as stewardesses. That rule made it less likely that there would be any shenanigans to distract crew from their jobs.

For quite a while, there was a ban on married couples with officer status working on the same ship. The officers' quarters were designed for men only, and there were no separate bathroom facilities for women other than for the captain and his wife, if he had one. Later the rule was dropped, and the company found that the married crew members were happier at their work if their spouse was aboard - that is if they both had a crew job.

My final lecture on North America's Great Railway Journeys drew a full house, 430 seats, and some were turned away. The projector had a new lens that allowed horizontal and vertical slides to be shown without any problem overlapping the dimensions of the screen.

On deck, while hugging the Baja California coast, I spotted spouting whales, playful dolphins, and my first leaping sailfish. The hot weather gradually changed to late fall conditions as we continued north.

The call at San Pedro (for Los Angeles) had the *Azure Seas* (former British-flag *Southern Cross*) and *Santa Marianna* in port, *Pacific Princess* in drydock, and the 1921-built *Princess Louise*, a former Canadian Pacific ship in Alaska service, now permanently tied up as a restaurant ship. A friend met me and we drove down the freeway to San Diego to ride the Tijuana Trolley that had just started the service at the end of July, south to the Mexican border and back. It left from outside the Santa Fe Railway-built Spanish Mission-style station.

On 1st December I arose at 6am to attend Cyril Smith's burial at sea. On the Poop Deck, a vast open

The author is standing on Oriana's monkey island with the golden cockerel behind indicating the fastest ship in the fleet. A small collection of the ship's souvenirs would sell out on this last voyage out to Australia. Oriana's small but distinctive funnel.

space, the crew was lined up facing aft and the flag on the stern flew at half-mast. The captain spoke into a microphone facing the crew with the deputy captain at his side. A deck hand held a floral wreath. The captain read the service, and a bugler played as Cyril's body went off the stern from B Deck along with a wreath thrown over the side. Just at that moment, the sun's disk appeared from behind the Coast Mountains, bringing a new day and brightening the sea and sky. The flag was raised to full staff, and everyone quietly dispersed. At the time, *Oriana* maintained her maximum service speed.

Cyril had kept a record of the ship's mileage sailed, a total of 2.8 million nautical miles. Nick Charlton, the officer of the watch, predicted that the ship would pass the 3 million mark in January.

Off the coast of San Francisco by 10am, the harbour pilot came on board, and we passed under the Golden Gate Bridge, while the ship's photographer hovered above in a helicopter. The photo must have been a glorious one, but I would never see it, as I would be disembarking here.

The docking was extremely difficult because of the narrow space between the finger piers and a strong tide running. Mike Moulin said that New York and San Francisco are the most difficult ports for manoeuvring, but they also have the best pilots. The P&O officers leave these dockings in their hands, while at others, the pilots are often humoured or even ignored.

The ship gently moved into the slip at a sharp angle with the bow heading almost directly toward the side of the pier before swinging parallel to the berth. Then I saw

Top: the Oriana climbs through the Panama Canal's three-chamber Gatun Locks to reach Gatun Lake.
Above: while the Oriana's Boat Deck did not offer a complete circumnavigation, nonetheless it was very popular with constitutional walkers.

my brother Sandy leaning out one of the pier windows and holding up a sign that read 'Welcome Ted Scull'. He had first gone to the Golden Gate Bridge but he was not allowed on the side facing the Pacific where I, and everybody else on deck, would be able to see the sign.

The next day, the two of us returned to the ship for a last visit and then with streamers tying the ship to shore, she went astern into the bay. Sandy and I ran to the car and drove quickly along the sea front to Fort Point, located under the Golden Gate Bridge's San Francisco side. We used the car headlights to flash 'Good Bye' and 'Ta Ta' in Morse code that one of the officers had written down for me. The response we got I could not read! *Oriana* then slipped under the bridge sounding a final whistle and headed out into the Pacific. Farewell *Oriana*; I had spent more days on that ship than any other.

My brother and I had a couple of beers while I talked about my idea of doing a book on line voyages. The first volume eventually happened 17 years later, and now in a series of three, here is the last, some 30 years later.

Oriana sailed from Sydney for another five years then in 1986 was sold to Japanese and Chinese interests for various stationary roles as hotel, restaurant, attraction and a museum, none particularly successful. At Dalian in northeast China, she was badly damaged in a 2004 storm and was dismantled in 2005. That was the last of the ships of the Orient Line, a company that started back in 1877 as the Orient Steam Navigation Company, although the origins of the business pre-dated that by 80 years.

Top: the Oriana at Acapulco. Bottom: the Oriana slides by the former Canadian Pacific coastal liner Princess Louise (built 1921), operating as a restaurant in Long Beach, California with the Azure Seas (former Southern Cross of 1955) in the background.

The Oriana, seen from San Francisco's Coit Tower, is berthed at Pier 35 on 2nd December 1981 before beginning her last trans-Pacific voyage to New Zealand and Australia.

Chapter Four

DFDS's SCANDINAVIA

Experimental East Coast Car Ferry Liner

6th - 10th January 1983

The venerable Danish company DFDS, dating from 1866, came up with the idea of operating a cruise ferry from New York south to Florida to attract motorists who balked at the long and exceedingly boring two- or even three-day drive. However, a major problem was that a non-U.S.-flag ship is not permitted to sail between two U.S. Ports without calling at a distant foreign port en route. Therefore, the ship could not directly link New York and Miami (1,330 driving miles or 2,140 kms) because a convenient call at Freeport or Nassau, in the Bahamas, did not qualify as distant.

The band-aid solution was for passengers and their vehicles to connect at Freeport to two of the firm's existing ships bound for Miami or to Port Canaveral (for Orlando), South and Central Florida respectively.

Factors in DFDS's favour were the collapse of the original privately operated Auto Train following the loss of passenger equipment and car carriers in two derailments and the potential market of three million motorists who drive annually between the Northeast and Florida.

The vessel that would provide the service, the MV *Scandinavia*, arrived in New York on 25th September 1982. With a gross tonnage of 26,747 and a length of 607 feet, the ship would leave New York every five days at 6pm for the 50-hour run to Freeport, where she arrived at 8am to connect with the 12:30pm departure of the *Scandinavian Sea* for Port Canaveral (arriving at 10:30pm) and the 5:30pm departure of the *Scandinavian Sun* to Miami (arriving at 11:45pm). Passengers paid for the second leg, but cars were carried free of charge.

Registered in the Bahamas, the *Scandinavia* had a crew of 354, comprising Danish officers with the rest being an international mix. As with some of the most modern overnight Scandinavian ferries operating in Europe, public rooms were located aft and cabins grouped forward to make sleeping conditions quieter. Of the 498 cabins, all with en suite facilities, 207 were inside, and the berth capacity, including upper berths, was 1600. Her vehicle decks could accommodate 530 mid-sized European cars.

I was eager to try out the new service and landed an assignment from Long Island's Newsday, the largest suburban daily newspaper in the U.S. The editor, who I knew well enough, said that there had been complaints about the new service so he stressed that and said "don't come back with a sugar-coated cruise piece". His only ship experience had been aboard a troop carrier in the 1950s, so he was no ship lover.

Scandinavian World Cruises' (DFDS) Scandinavia arrives New York for the first time on 25th April 1982.

Above: shuffleboard on the Scandinavia in breezy January conditions. Below: the swimming pool aboard the Scandinavia is housed under a Magrodome.

On 6th January 1983, about three months into the *Scandinavia's* service, I booked a round trip. Then, on the day of sailing, word came from Scandinavian World Cruises in Miami that due to a severe Atlantic storm, the *Scandinavia* would not sail on Wednesday night as scheduled, but would embark on Thursday at 8am and sail by 10am. The bonus for the delay was a TV showing of 'Ship of Fools', a favourite film that I could now watch before leaving home.

After an early rise at 6am, two connecting buses brought me to North River Pier 92 where the *Scandinavia* was berthed, stern in to be able to handle vehicles via a newly constructed ramp. However, instead of boarding at 8am, we embarked at 10:30 and sailed at 11:15, nearly 12 hours late. The New York turnaround was accomplished in a speedy four hours 15 minutes, but no one on the pier or on the ship quite knew the expected arrival time in Freeport. That uncertainty would begin to rattle the motorists.

Travelling with a companion of Scandinavian descent, we occupied Cabin 643, amidships on Atlantic Deck, where the restaurant is located. Two beds were placed either side of a large window, and flexibility allowed the cabin to become a sitting room during the day. The bathroom was tiny, and the TV showed only the daily programme and announcements.

The restaurant - Windows of the World - was bright and cheerful with huge windows looking out to the sea. Around 700 passengers could be accommodated at each of two sittings. Our table companions were a couple from Babylon, on Long Island, he a former printer and now an antique dealer and she a librarian at a junior college, and a couple from Montreal, he making uniforms for the Royal Canadian Mounted Police and she a housewife. From the initial conversations, I felt we had the makings of a great table.

Public rooms were all positioned at the stern, one above the other. On the highest deck were the Galaxy Night Club and the Blue Riband Club; the latter dedicated to the great liners and displaying black and

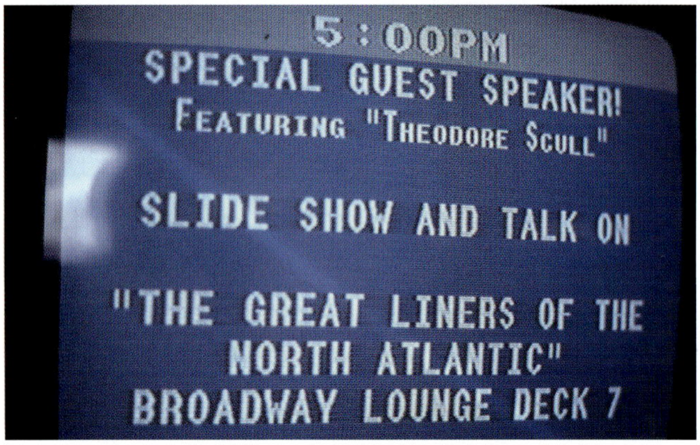

The author, bundled up during a January sailing aboard the Scandinavia from New York to Freeport is tempted to ring the bell - but decides otherwise, and just for the fun of it, the author gave a slide show to the motorists aboard the Scandinavia.

white photographs and two reproduction paintings of the *United States* and *Mauretania*, both Blue Riband winners. Below that was the Broadway Lounge, by far the largest room on the ship, with the casino on the same deck offering slot machines, blackjack, and roulette tables. The sole intimate room on the ship was the attractive John Paul Jones Pub furnished with wooden chairs and cosy booths.

The swimming pool area reminded me of a health club at the roof level of a condominium; glass-enclosed with a sliding roof that remained in the closed position. The high noise level produced an unpleasant echo of voices reverberating off the hard surfaces. The space with its bar, while visually attractive, lacked anything like a relaxing atmosphere. On 7th January, it was slightly warmer, but not yet warm enough for the Magrodome to be pulled back.

The generous outdoor space provided ample walking paths on several levels and plenty of deck chairs. The decking was synthetic green matting and had become quite sodden after all the rain.

The funnel loomed tall and massive, located just in from and above the built-up, squared off stern. While standing on the Signal Deck facing forward, the stubby bow was invisible until one reached the railing and looked straight down. The superstructure, carried so far forward, resulted in practically no traditional bow area at all, and what was there became badly crumpled during a storm on the very first trip and remained that way during her New York service period. Overall, the impression was positive, and nothing about the ship seemed cheap looking.

Passengers were mostly older, accompanied by very few children, and definitely motorists, not one-way or round-trip cruise passengers, as the car deck was full and the passenger count 750. As the ship's berth capacity was 1,600, DFDS would have to attract a lot of round-trip foot passengers to fill it. Thus far, the largest number the ship had carried was 1,100, and the least, 300 on a northbound sailing. I already had my doubts how long this would last.

Prior to the start of the voyage, the cruise director had given a talk about landing in Freeport and had created more problems than he solved, partly because he was reading the script as if we were arriving on time at 8am, including the request to have bags placed outside the cabin doors at 1am in the wee hours of the night before.

Apart from the confusion, many passengers were angry about having to pay for hotel expenses in New York, and some shouted their questions. With a scheduled evening sailing, motorists expected to simply drive directly to the pier and not have to deal with anything New York at all, let alone, hotel, meals and parking, and whatever else might befall them. After all, this was New York City in the 1980s.

Another problem that arose would be the delayed sailing of the *Scandinavian Sea* (ex-*Blenheim*) to Port Canaveral, which would now become an overnight passage. The cruise director could not tell the passengers if there would be cabin accommodation available unless they phoned ahead. When asked to phone ahead for them, he refused point blank.

The *Scandinavian Sun* (ex-*Freeport*, *Caribe*, *Svea Star*) would, however, sail on time at 5:30pm for Miami. The awkward connections even at the best of times just might kill the service.

The litany of confusion continued. The baggage tags and their colours proved endlessly confusing to passengers because there were so many choices: 1) transfer immediately to one or the other ship, 2) stay over in Freeport, 3) park the car in a bonded area or 4) secure a permit to drive in Grand Bahama, for a fee.

Happily for some, I had offered to give a slide talk about the Atlantic liners in the Blue Riband Club. About 200 people showed up including the captain and first officer. Several came up afterwards to say that they had sailed in the Cunard Queens during the war or on a number of other ships.

On 8th January, the captain, in an interview in the Blue Riband Lounge, reported perhaps rather too fully to his passengers that that the company (DFDS) was having a very hard time financially because of start up costs, and that the service was losing a great deal of money. The Mediterranean ferry services that DFDS started had done very poorly and had been closed down in the preceding few weeks. The North Sea ship *England* had been sold to the British Government for trooping - a post-Falklands War measure. DFDS had taken over Prins Line operating Harwich-Bremerhaven, and the route was then shut down, leaving the Hamburg service with just one ship operating every other day. Currently, there was a strike in Denmark's Esbjerg Harbour that had stopped the important ferry service to England.

I don't think that I am out of line by saying that I may have been the only passenger in the lounge who had any idea what he was talking about, other than that DFDS was in deep financial trouble. I can only imagine the turmoil at the home office in Copenhagen. The passengers now on board were worried about their own fate here, not some impending corporate upheaval thousands of miles away.

I spoke to the captain after the talk mentioning that my first overnight ferry had been DFDS's 1937-built *Kronprins Olav* between Copenhagen and Oslo in

The Scandinavia arrives Freeport, the Bahamas on 8th January 1983 to connect with the Scandinavian Sea to Port Canaveral and Scandinavian Sun to Miami.

July 1960. From 1965 the ship operated on the much more open-water Denmark to Iceland route, but the captain said she was a top-heavy ship and especially unseaworthy on that run, so had been sold for service in the Mediterranean just two years later.

At lunchtime I was on deck to watch the Freeport arrival, with lots of tankers at anchor and more at the oil jetty. After passing through the narrow entrance into the harbour, the *Sun* and *Sea* came into view, bringing the whole fleet together at one time. The *Scandinavia* went astern into the slip alongside the *Sun,* which was offloading its passengers and had its bow up ready to receive cars from our ship.

Cars and passengers began disembarking within 15 minutes of our 2:30pm arrival. Some drove directly onto the *Sun*. I disembarked and walked over to board the *Sea.* She seemed dowdy and much smaller by comparison, and she had very low ceilings. While the dining room was attractive, the lounges were filled with slot machines and the once attractive forward lounge was now a casino. But then its main market was shuttling people to the Freeport casinos, with few pretensions as a cruise-style ferry.

The Scandinavia loads cars at Freeport on 8th January 1983.

I met Captain David Keighley, an ex-British India man, who had served in the Indian Ocean aboard the passenger vessel *Sangola*, the troopship (and later school ship) *Nevasa* and several cargo ships. He had had four years on the *Sea* when she was Fred. Olsen's North Sea ferry *Blenheim*, and when he came over to DFDS he maintained his command because the ship was registered in the Bahamas. What a contrast of ships and what a comedown in importance.

The *Sun* left at 5:30pm, and we followed right behind. Back on the *Scandinavia*, the return table with new people was beyond painful, as this group did not talk at meals beyond "pass the salt conversation".

On the morning of 9th January, the ship experienced a two-hour power failure, apparently a problem with the oil filter and the auxiliary diesels. Everything shut down - lights, air conditioning, power to pump water and flush toilets, and to operate the propellers. We wallowed around in relatively calm seas with, from time to time, the engine turning over briefly and the funnel producing a plume of black smoke. We got underway again at 11:15am. Walking about the ship, I noticed that many men were sitting in chairs with their arms crossed wondering, no doubt, "When will this voyage ever end?"

My second lecture "Sea Voyages around the World" could be considered an introduction to sea travel for the motorist, being the development of the car ferry in Europe, cruise ships in New York, Miami and the Bahamas, Alaska Marine Highway, Greek island ferries and cruise ships and long-distance liners. The captain and staff captain again attended along with about 200 others. Small clusters got talking afterwards.

We hit thunder, lightning and very strong winds when rounding Cape Hatteras, and what a dramatic sight watching the spray whip past the huge windows and the tempestuous seascape illuminated by the intense lightning bursts. It was a wild and windy walk after dinner, and the pounding seas caused loud reverberations in the huge cavity that housed the vehicles. It was difficult getting to, and then staying asleep.

On 10th January, I awoke to a grey, wet and windy morning that progressively got worse resulting in an almost invisible arrival back in New York with the ship enveloped fog and driving rain. We saw nothing until the ship was making its turn just off Pier 92.

I did not have any problem avoiding a sugar-coated cruise story, rather how could it be made at all appealing for a market made up of largely motorists who only wanted to avoid the intensely uninteresting two long days of driving. This service was simply not to be, and

The Scandinavian Sun is leaving Freeport for Miami on 8th January 1983. The Scandinavia plows northward to New York through heavy seas on 10th January 1983.

with all the other problems DFDS was having, the *Scandinavia's* operation almost plunged this venerable firm into bankruptcy. By 1984, after barely two years in service, she was withdrawn.

In 1984 the ship briefly transferred to the DFDS overnight service between Copenhagen and Oslo, and was then sold to Sundance Cruises as the *Stardancer*, still taking vehicles on West Coast routes. In 1990, Royal Caribbean took her on as *Viking Serenade* for short cruises from southern California to Mexico, now rebuilt with cabins added in the former vehicle space. Finally, in 2002 she moved to European waters to operate as the *Island Escape* for Island Cruises until 2009 when she began operating seasonally for Thomson Cruises in the Mediterranean from March to October. Withdrawn in 2014, at the time of writing she is serving as a refugee accommodation ship in Norway under the name *Ocean Gala*.

Happily for those making the annual trek between the Northeast and Florida, the Auto Train returned, this time run by Amtrak. I was aboard for a first trip, again for Newsday, and it was a much easier story to write. The motorists were over the moon to have their train back, and the operation is still going strong three decades later.

Chapter Five

P&Os CANBERRA

A Line Voyage from Australia to England

1st March - 10th April 1985

In May 1852, the fledging Peninsular and Oriental Steam Navigation Company sent its diminutive 699-ton sail-assisted steamer *Chusan* on a pioneering 77-day voyage from Southampton to Melbourne, via the Cape, marking the beginning of a long association with the United Kingdom-Australia passenger trade.

Just a couple of months short of 133 years later, on 1st March 1985 at 11:07am, P&O's 44,807-ton *Canberra*, the largest liner ever built for this route by any company, slipped her moorings and eased away from Sydney's Circular Quay to begin a 40-day homebound voyage to Southampton via Suez.

With Captain David Hannah and Deputy Captain Ian Gibb in command, *Canberra* carried a full complement of passengers - 1,113 Australians, 407 British, 78 New Zealanders and a scattering of thirteen other nationalities, including at least one American.

For many years prior to this, both *Canberra* and *Oriana* had voyaged to and from Australia via North America and the Panama Canal, so the return to the traditional Suez Route made for a welcome change.

The homeward voyage had three distinctive sections - an old-fashioned interstate run from Sydney via Melbourne and Adelaide to Fremantle, a traditional liner run from Fremantle to Suez, with Bali thrown in as a cruise port, and a multiport cruise itinerary in the Mediterranean and on to Southampton.

The last section appealed to Aussies and Kiwis who were bound for an extended holiday in Europe, while the interstate portion revived what was once a popular short voyage, and the demand exceeded the space available by a wide margin.

Autumn started on 1st March in the Southern Hemisphere, and Sydney's weather could not have been more co-operative. On this crystal clear day, the *Canberra* looked absolutely splendid alongside Circular Quay. The waterfront swarmed with activity as several thousand passengers, their relatives and friends made their way to the telescopic gangways for embarkation. Heaps of personal effects were transferred from shore to ship for distribution to the cabins, baggage room and the hold.

The Canberra at Circular Quay on sailing day for Southampton on 1st March 1985.

Bon Voyage parties filled the public rooms and cabins. For those occupying the tightly-packed accommodation aft, the celebrations spilled out into the corridors, tangling with the bewildered newly embarked searching for their cabins and blocking stewards trying to deliver mounds of baggage.

At 10am, the officer of the watch commenced a string of announcements requesting visitors to proceed ashore. The warning took on special importance when he added the phrase "now, by the final gangway".

Passengers began drifting to the open decks to link up again with those not sailing via a thick tangle of coloured paper streamers. A little after 11am the last lines were released, and the *Canberra* slowly edged away from the pier. With three blasts from her whistle, she went astern into the main harbour channel, her

Top left: the Canberra begins to move at the start of the long voyage back to England. (Photo: Bruce Miller) Top right: Sydney's Circular Quay gave one of the best sendoffs in the world, with several thousand visitors coming aboard at embarkation.
Above: the Canberra sails from Sydney's Circular Quay on 1st March 1985 on a voyage back to England. (Photo: Bruce Miller).

A peaceful, sunny afternoon aboard the Canberra lounging under the boats.

stern stopping just short of the Sydney Harbour Bridge. Tugs positioned her, and the bridge rang 'Slow Ahead'; Sydney's dazzling opera house gradually receded to starboard, and *Canberra* made for the Heads and the open sea at the start of the 14,208-nautical-mile Voyage 561 back to Britain.

On the following day, *Canberra* pitched gently into a Force 8 while steaming through the island-studded Bass Strait. Some passengers had not taken kindly to the choppy sea conditions, though by mid-afternoon, they got their relief when the Melbourne pilot came aboard, and the ship crossed the 'The Rip' to enter Port Philip Bay and berth at the Station Pier by dinnertime.

The *Canberra* had not called at Port Melbourne since 1972, and a huge crowd had gathered hoping to get aboard. In the end, some estimated 2,500 visitors thronged the ship until well after midnight, and calm was not regained until 2am when the ship sailed for Adelaide.

Early on 4th March, *Canberra* cleared Backstairs Passage entering the Gulf of St. Vincent to dock at Adelaide's Outer Harbour Terminal just before 8am. Here visitors, merely numbering in the hundreds, included good friends John and Sandy Roe who lived nearby on Aurelia Drive, one of many streets named after passenger ships. Other nearby streets were named *Ellinis*, *Himalaya* and *Oronsay*; ships that brought migrants to the country.

On 7th March, *Canberra* passed Rottnest Island and entered Fremantle's long harbour basin. Thankfully, the ship docked upwind of Kuwait-registered *Al Shuwaikh*, the world's largest livestock carrier, which was in the process of loading an incredible 130,000 sheep for transportation to the Gulf. Lorry after lorry arrived and, hour after hour, a steady stream of animals flowed up the ramp into the ship. Fremantle had traditionally been the first and last port of call for ships trading between Europe and Australia, until regular voyages ceased just under a decade earlier.

Upon departure, Western Australia's Police Band played on past their planned finalé when immigration formalities delayed the sailing. The extra time turned out to be a stroke of luck, for visiting friends Alistair and Pam Mees had left their pram in my cabin, and it took me nearly all that time to manoeuvre it through the crowds and off the ship.

In the gathering darkness, *Canberra* was now firmly northbound with almost the entire complement of passengers bound for Britain. The line voyage had begun in earnest. One significant sign was the sight of hundreds of passengers doing their constitutional around the open promenade deck; five circuits equalled a mile. Many other passengers gathered at the stern to feel the rumble of the propellers under foot and to watch the wide wake boil astern.

The *Canberra* offered the widest range of accommodations afloat, allowing the middle class and millionaires to enjoy sea travel in the same boat. Her interior design reflected the original two-class configuration. There were inside four-berth cabins, with shower facilities down the passage on 'G' Deck, spacious verandah suites with separate sitting areas and 32 more cabin grades in between.

I occupied a court cabin, for single occupancy, with corner windows that looked into the court and beyond to the windows at the side of the ship, bringing in natural light and affording a slightly removed view of the sea. The court's sofa under the windows became a gathering place and the coffee tables a repository for excess bon voyage floral bouquets.

Once, first-class passengers lived in the forward part of the ship and dined in the grand and spacious Pacific Restaurant, and tourist-class passengers, located aft, in the more densely packed Atlantic Restaurant. As a one-class ship, cabins 1 to 103 were assigned to the former and cabins 201-333 to the latter. Apart from an additional appetizer and entrée, the menus, if not the atmosphere, were much the same. Two sittings were scheduled for all meals, and at dinner when the ship was at sea, ship's officers, right down to the third rank, hosted tables in both restaurants. Passengers loved to talk about 'their' officer.

The restaurant and cabin stewards were European, mainly British, and Goanese Indians, often Christian, from what was Portuguese Goa, a colonial enclave on India's west coast. The deck hands were Muslim Indian; the engine crew was Muslim Pakistani; the ship's carpenters Chinese, and the officers British.

P&O cooking was strictly British with a wide selection at every meal. Some passengers, like me, cannot refuse the lunchtime daily curry with all the condiments. At dinner, the roasts were excellent, and the fabulous cheese board offered a changing selection of a half-dozen varieties, plus a crock of Stilton, to match with eight different kinds of biscuits.

The variety of public rooms was so extensive that everyone could find a favourite spot during the day. 'Canberra News', the daily programme broadsheet, offered so many activities that it took most of breakfast to mull over the options. Barbara and Charles were giving a lecture on pewter design in the William Fawcett Room; Derek Rimington was hosting a bridge lesson in the card room; and Syd and Edna were teaching the two-step in the Bonito Club. Elwyn Griffiths would have three sessions for his talk on Bali, the next port of call, and regardless of the local geography, he ended every lecture with a Welsh sunset. The passengers loved him. All these entertainers and lecturers had been with the line for years and in some cases decades. I had been with P&O for six years, coming aboard several times a year, and happily I got on with these characters quite well, as long as my talks were not scheduled at the same time as theirs.

Deputy Captain Ian Gibb aboard the Canberra radios a navigation order to the helmsman. He would rise to become Commodore of the P&O fleet; one of the bridge telegraphs.

Above: two guys go at it on the greasy pole above the Canberra's Bonito Pool. One of the decorative signs by the Bonito Pool.
Below: the Cricketers' Tavern (photo: William Mayes).

Deck tournaments included deck tennis and table tennis for women, men and mixed; passengers vs officers cricket matches; the occasional tugs of war; and lively pool games. Many who weren't taking part watched with great interest, and yes, lots retreated to deck chairs located near one of the pools, on the forward observation deck often under the broiling sun, along the sheltered open promenade or a quiet out-of-the-way recess on B Deck aft.

As evening fell, there were no less than nine locales to have a before dinner drink or two. They were highly varied, from the pub atmosphere of the Cricketers Tavern, decorated with sporting murals and bats and hats; the forward-facing wraparound Crow's Nest with a pianist and the best views of the sea; the tiny tucked away Century Bar and the formal, red plush Meridian Room. Few passengers dressed formally on a line voyage where the norm for men was a suit or at least a jacket and tie.

After dinner, many passengers liked walking the open decks before they headed for the cinema, a cabaret act in the William Fawcett Room, the Broadway-style show in the Stadium, quiet dancing in the Bonito Club or a classical concert in the Meridian Room that often devolved into a bit of a sleeping lounge during the performance. Every evening the Syndicate Quiz in the

Peacock Room drew more than a dozen tables of six and became very competitive as the voyage progressed. I hardly ever missed the event.

During lunch on 10th March, Deputy Captain Ian Gibb came over the tannoy requesting my immediate presence on the bridge. With the lifts far too slow, I bounded up 138 steps arriving just in time to see a passenger ship closing with us at a combined speed of almost 40 knots. After a bit of a panic, I was able to identify the ship as China Navigation's *Coral Princess*, built in 1962 for Brazilian coastal service as the *Princesa Leopoldina* and bound for Fremantle. *Canberra's* hugely impressive whistle was answered by a decidedly weaker but no less friendly response. Feeling rather proud of myself having passed muster with the deputy captain, I returned triumphant to my curry lunch.

The approach to Bali through Lombok Strait was serene and tropical early on the following morning. The *Canberra* dropped anchor in Padang Bay beneath the volcanic peak of Mount Agung. The sacred 10,000-foot mountain had last blown its top in 1963 causing considerable destruction to life and property.

Outrigger canoes gathered around *Canberra's* stern carrying Balinese salesmen enticing passengers to purchase their wood carvings and coaxing the Indian and Pakistani crews to buy live chickens. Endless wrangling continued even after the ship had weighed anchor,

The Canberra is sailing due west through the Indian Ocean bound for Suez, and the author at the Canberra's railing on a tropical night in the Indian Ocean.

producing a few now desperate additional sales. Our day ashore had been spent in drenching rain.

The Crossing of the Line ceremony took place around the Bonito Pool on the afternoon of 12th March while the ship steamed through the Java Sea at more than 23 knots. The outdoor setting is much like an ancient world amphitheatre, and the treatment that the cruise director Phil Raymond as King Neptune ordered for his poor victims, including Deputy Captain Ian Gibb, recalled the messy tortures the Romans meted out to the Christians. In the modern-day scenario, the wretched souls were thrown to the dressed up bears, in the ship's pool, rather than to the lions. It's not only the victims that suffer Neptune's wrath. The Indian crew had to mop up the considerable mess created by all the spilled 'blood' (red confectioners' sugar).

Navigator Trevor Lane, in his daily noontime announcement of the ship's position, speed, weather and the like, enquired if any passengers had felt the slight bump at 0715, the moment when *Canberra* actually crossed the Equator! Colourful certificates, signed by Captain D.H. Hannah, were delivered to every cabin.

Lots of shipping of all types and sizes appeared over the bow as *Canberra* entered the busy shipping lanes on approach to Singapore, a shipping enthusiast's dream. Hundreds of ships crowded the Eastern and Western Anchorages, but since I made my first visit here in 1968 and again in 1974, many colonial-era buildings had been demolished and the transformation had produced a new world-class city skyline.

The *Canberra* docked at the brand-new western container berths for the next two nights and a day, and I explored the city, waterfront and Singapore River looking for the bones of the city I once knew and learning to accept progress. Upon departure, the ship soon entered the Strait of Malacca, one of the world's busiest shipping lanes with two convoys heading in opposite directions.

At the beginning of a double lecture session in the theatre on 16th March, I asked my audiences if they knew what was special about the following day. Three answers burst forth: Mother's Day (UK), St. Patrick's Day and the loudest response '*Canberra's* 25th birthday', the answer I was looking for.

On 17th March 1960, Dame Pattie Menzies, wife of the then Australian Prime Minister, Sir Robert Gordon Menzies, had christened the ship at Harland

Setting up a Canberra deck chair led to angst and laughter. Let's start all over again. Both the Canberra and Oriana had at least one deck chair that was impossible to set up.

and Wolff, Belfast. Dame Pattie's portrait was found in the forward foyer on Promenade Deck. On the following day, I opened the lecture with 'Happy Birthday *Canberra.*' and an enthusiastic round of applause ensued. Many *Canberra* passengers, as it turned out, were latent ship enthusiasts.

At dinner that night, Captain Hannah read a telegram from Dame Pattie, gave a toast, and ordered Champagne for all. With the lights out, a special birthday cake was paraded through the Atlantic and Pacific restaurants.

Colombo's old-fashioned harbour facilities made a welcome change from the stark efficiency of modern container ports. Most passenger ships making runs to the Far East and Oceania called here, and until about 15 years earlier, they brought the lion's share of the tourist revenue. Sri Lanka's main port today had an air of time standing still. Wooden launches of every description brought out officials of any government department that could find an excuse to pay *Canberra* a visit and perhaps be invited to a meal and take away a gift.

While there were two other passenger ships in port today, the Russian *Taras Shevchenko* and *Odessa*, P&O had first refusal on the Sri Lankan Railways' special excursion train, and 100 passengers, including myself as escort, made the wonderful three and a half hour climb into the hills to Kandy, the country's Buddhist centre and last capital of its ancient line of kings. The day offered splendid mountain scenery, the famous botanical gardens at Peradeniya and the Temple of the (Buddha's) Tooth. While armed guards rode the train, most of the trouble was, at the time, confined to the north of the island in the Tamil-speaking areas.

After Colombo, there were four restful days at sea crossing the Indian Ocean. At sundown on 19th March we passed two miles south of Minicoy Island in the Indian Laccadives. A large lighthouse sent its sweeping beam seaward while a glorious orange-red sun dropped straight down *Canberra's* forward flagpole fixed to her bow. Nearly everyone in the Crows Nest was drawn to the windows, and there was no doubt that we were headed precisely due west.

On 22nd March we rounded the Horn of Africa off Cape Guardafui, a backdrop of jagged, dry mountains, to enter the Gulf of Aden. Few ships now call at the port of Aden for bunkering, following the abrupt change in the political climate. Instead, the ship made for Djibouti, where we arrived on the following morning.

Formerly French, steamy Djibouti was a scruffy sort of place without much colonial charm, but with plenty of local colour, especially at the African market

with so many different peoples engaged in buying and selling. The port served as the terminus for the railway line inland to Addis Ababa. At the station, a goods train, carrying food supplies to the starving population, was preparing to leave for the capital.

Shortly after noon, with bunkering finished, we aimed for the Red Sea. On the next day, the air was filled with fine sand caused by the strong prevailing headwinds. That night, some passengers gathered on the forward observation deck, while the deputy captain, using a powerful torch picked out the constellations. In the hushed silence of the dead calm, with no perceptible movement of the ship, it appeared we were on an intergalactic mission, suspended in space. The low-lying Southern Cross, once high in the sky, indicated that we were fully in the Northern Hemisphere again.

Now in the Gulf of Aden, both the Egyptian coast and Sinai Peninsula were clearly visible. Then two mornings later at 5am, anchoring off Suez amongst several dozen other ships, the first of 860 passengers began disembarking for the long day excursion to Cairo and the Pyramids.

Canberra got under way at 6:24am and fell in behind Hapag Lloyd's *Tokio Express* with a pace that kept the two ships ten minutes apart. The pilot, the most senior for the

Suez Canal Authority came aboard dressed in a flannel suit and felt cap. Before he got down to work, he made his first requests of the day - a stool, breakfast menu, a list of the types of cigarettes and cigars carried on board and a large bottle of whisky. There was a hushed silence as the deputy captain responded by saying that there was neither a menu nor a list. Instead he was offered an egg sandwich, the de rigueur early morning sustenance on the bridge. When he was handed Johnny Walker Red Label, he turned it down and demanded Black Label. One could have cut the air with a knife when the deputy captain asked, 'Will that be all'?

While the telegraph remained on 'Half' for most of the morning, the pilot constantly ordered changes in the revolutions per minute from 45 to 50 rpms. The navigator kept him informed of the cable lengths (608 feet), and if the gap narrowed to five, he ordered 'Dead Slow'.

Except for passing points, the traffic advanced single file. While there was some evidence of military presence in army camps and watchtowers, most damage from the Arab-Israeli fighting was no longer evident, apart of the occasional ruined building. The view was mostly of wide expanses of sandy desert on both sides.

In the Great Bitter Lake, we were held for nearly an hour, and when some of the pilot's demands for 'baksheesh' were not met to his satisfaction, he, quite openly, threatened to delay the ship by four hours.

At Ismailia, the half waypoint, there was a change of pilots. The replacement was the deputy director of the Suez Canal Authority, and a gentleman professional. He called for a course of 0.00 or due north, and the normal relaxed atmosphere soon returned.

On the approach to Port Said, the triple-domed Suez Canal Authority headquarters appeared to port. Now under the control of a newly-embarked docking pilot, he raised a few eyebrows when he spat a few times on the bridge wing while supervising the tricky manoeuvres in a very tight space. Once tied up, the snake pontoons uncurled to form a floating S-curve gangway between ship and shore.

The huge ship towered over the modest Port Said skyline, and soon small boats, packed to the gunwales with leather goods, woodcarvings and brass, vied for a position along the length of the snake. The arcaded shops in the adjacent streets gave Port Said an inviting and timeless look.

Passengers returned from the Cairo excursion at dinnertime, having had a wonderful if exhausting day. *Canberra*, due to sail at 9pm, was held up until 10:15, when the docking pilot did not appear on time, a last swipe at the ship that had not shown enough hospitality.

The departure was painfully slow, and once free of the pilot, *Canberra* revved up to her top speed to make Haifa, where there would be nearly forty coaches

Mahmood, Indian deck serang, had charge of the deck crew for many years, and on every voyage he invited me to his cabin for a chat and a lunchtime curry cooked by the crew's chef.

waiting to take passengers on full-day excursions, as close to time as possible.

The line voyage section was now at an end, and the Mediterranean cruise had begun. With seven ports in the next eleven days - Haifa, Izmir, Piraeus, Naples, Marseille, Gibraltar and Vigo, the gentle rhythm had vanished, and we received an overdose of significant sightseeing. Many passengers became extremely tired and *Canberra's* infamous 'cough' laid low quite a few, with some really quite ill. Attendance at lectures and activities dropped significantly. We were now looking for the voyage to end.

Easter Sunday, the sixth consecutive Sunday at sea, a packed congregation in the Stadium sang Hymn 370 'Eternal Father strong to save…For those in peril on the sea' for one last time, and with a little extra gusto.

Upon departure from Vigo, a warning was broadcast about nasty weather in the notorious Bay of Biscay, but all the Bay could muster was a paltry Force 4. By nightfall on the eve of arrival back in Southampton, *Canberra's* corridors and alleyways were piled high with suitcases, signalling without question, the grand finale.

The largest group of the voyage packed the Bonito Club to hear the popular D-Amour duo play favourite dance tunes for the last time. Nobody wanted to go to bed, but soon the Southampton pilot would be boarding to take the ship up the Solent and into Southampton Water.

At 7am, right on schedule, the telegraph rang 'Stop' engines, and with the *Canberra* alongside berth 106, Voyage 561, and one of the last extended point-to-point voyages anywhere in the world, was at an end.

Canberra remained remarkably popular as a cruise ship right up to her withdrawal when in September 1997 she left Southampton under the cover of darkness to be broken up at Gadani Beach, Pakistan. The scrappers, who expected to demolish her in three months, ended up struggling with her sound construction for almost a year, losing money in the long process.

Canberra, a ship that had played a major role in the Falklands War, remained resistant to her demise right to the end, and her impact on hundreds of thousands of people's lives had been immense.

Chapter Six

Turkish Maritime's AKDENIZ

Last Turkish Coastal Liner

5th - 15th August 1987

Turkish Maritime Lines once operated one of the most beautiful passenger ship fleets in the world; a particularly handsome set of cargo-passenger liners largely built in Germany and Italy in the 1950s along with highly varied second-hand tonnage from the United States. Regular services radiated from Istanbul along the Turkish Black Sea and Mediterranean coasts and to Eastern and West Mediterranean ports from Beirut, Haifa and Alexandria to Piraeus, Venice, Naples, Genoa, Marseille, and Barcelona.

I had first seen a number of these intriguing ships in 1969 while traveling aboard TML's *Izmir* on a short voyage from Rhodes to Istanbul and a week later aboard the Adriatica Line's *San Marco* from Istanbul via Piraeus to Venice and Trieste. A proper voyage was high on my wish list.

By the time I was able to make such a trip, much of the classic and second-hand fleet had been retired and the choices were few. The 1955-built *Akdeniz* that I had seen in Naples some years earlier, was now operating on a coastal run from Istanbul to Turkish ports along the west and south coasts as far as Alanya.

Early in 1987, I wrote a letter to the head office in Istanbul about a passage and received a courtly typed reply a few weeks later with apologies that the ship was fully booked.

Suellyn, then my girlfriend and four years later to become my wife, knew a Turkish student at her school, and she asked her to come to her office. Her father had been high up in the Turkish government so we inquired if he might help us to get aboard. She said that she would try, phoned her father and then came back saying that it might take some time.

Roughly eight weeks later, a phone reply from her father advised that we had a cabin for 6th August, and we could pick up our ticket upon arrival in Istanbul. I was thrilled and straight away launched into creating a summer's European itinerary.

On 29th July, I boarded a train in Gravesend, Kent, changed at Chatham and met Suellyn who had

The Akdeniz is docked at Alanya on Turkey's south coast.

come down from London at Dover Western Docks. We embarked in Belgian Marine's *Prince Laurent* for the 3pm departure, and during the crossing, enjoyed an early three-course supper in the small waiter-service restaurant.

Arriving Ostend at 7:30pm, we headed for Train 299 to Basel via Brussels, Luxembourg, and Strasburg. The sleeping car attendant looked at our ticket, which showed the routing via Aachen, Cologne and Basel and informed us that it would not be valid for this train. So I dashed into the ticket office, asked if I could jump the queue and had the ticket changed with $28 returned to each of us. The attendant took our passports and tickets, and we filled out currency and police forms for France. Our T2 sleeper had nine upper and lower berth compartments. The train departed at 8:53pm with a light load, then took on many more passengers at Brussels Midi. After a comfortable night, the sleeper was gently shunted at Basel to another train, taking us along a twisting Swiss route of pretty towns, tall church spires, rushing streams and several tunnels, including the 9-mile (15 km) St. Gotthard. On the far side, in Italy, it was raining as we skirted Lakes Lugano and Como. Italian passport inspection took place at Chiasso.

Arriving Milano Centrale on time, we sprinted from Track 4 to 18 for an earlier, faster train, to Lecce and climbed aboard the last coach just as the door was being shut. The conductor indicated that this was not the train on our ticket but gave us the okay, charged us the supplement and then said there were no seats. We camped on the carpeted floor at the end of the train looking back over the tracks. The train went express at 140-160 km/h via Bologna and hit the Adriatic coast at Rimini. Disembarking at Ancona, we walked three kilometres to the maritime station, having been told that it was just one.

The *Queen Vergina*, registered in Limassol, had recently been bought from the Japanese and revealed her East Asian origins on the bow and the name board attached to the bridge wing. As the ship was ultimately bound for Israel, boarding was an excruciatingly long process: immigration, police and security check for baggage, then handing over passports at reception for some 450 passengers, 77 vehicles and 13 motorcycles.

Our spacious air-conditioned cabin had twin beds, sitting area, two very tall windows dropping to within a foot of the floor, heavily stained fabrics and rice paper wall coverings. Meals were extra and paid for in advance at the reception. As the restaurant air-conditioning had failed, we ate the non-descript meal quickly and fled out to the deck for air.

Breakfast was yucky juice, burnt coffee, stale bread, jam, biscuits, and one semi-hard-boiled egg. Other meals were slightly better, probably because we were hungry, and otherwise, we spent most of the day in canvas chairs on deck.

The ship called at Corfu and Patras and then after transiting the Gulfs of Patras and Corinth, entered the Corinth Canal with a pilot and tug assisting for the tight squeeze. It was a hot two and a half hour transit between

The Queen Vergina, freshly purchased second hand from Japan loads passengers at Ancona, Italy for Patras, Piraeus, Limassol and Haifa.

the high rock walls. We arrived in Piraeus at 3:30pm, and when immigration did not show, we went to the purser, collected passports and walked off.

After a two-night stay at the shabby Nofara Hotel, although happily with air-conditioning, we sailed in Sun Line's *Stella Oceanis* on the evening of 3rd August with a call the next morning at Dikili and a bus tour to Pergamum. The ship then entered the Dardanelles and on into the Sea of Marmara arriving at Istanbul early the second morning at 7am.

The *Akdeniz* gradually appeared through the fog alongside the pier, its spotless white hull and mustard-co-

Top: Sun Line's Stella Oceanis en route from Piraeus to Istanbul stops at Dikili, Turkey.
Above: a week after leaving London we arrive at Istanbul to transfer to Turkish Maritimes Lines' Akdeniz for a voyage to the South Coast.

loured masts and funnel set smartly against the drab docklands.

Docking just astern, we were off at 8:15 and hustled over to the stately Turkish Maritime Lines' headquarters to collect our passage ticket and were told to come back at noon. Returning 15 minutes before that, the one clerk on duty in the vast booking hall began opening drawers, and after checking about a half dozen - voila! There it lay in an otherwise empty drawer, and I started breathing normally again.

At embarkation early that afternoon, the quayside was thronged with passengers, relatives and friends, and all manner of impedimenta. Once aboard, we entered another world; the era of the 1950s with decorative throwbacks to the period before the Second World War.

In the purser's foyer, flying saucer-shaped lamps jutted out from angled metal brackets attached to beautifully wood-panelled walls. In one of the two lounges, afternoon tea was being served with piano accompaniment. On this same deck, our accommodation, one of 16 outside cabins with private facilities, was decorated with dark woods and had an old-fashioned chaise and easy chair upholstered in rich green velour. I was thrilled and threw my arms around Suellyn who then said very little which, in fact, was a lot. One of us was overwhelmed and the other quite the reverse.

Cemil, our delightful moustached cabin steward, proved to be the principal link of communication to everything and everyone. He had been with the TML for 26 years and had worked aboard the 1927-built *Ankara, Iskenderun, Karadeniz and Akdeniz*, and he really enjoyed having English-speaking passengers in his charge.

At 2pm the *Akdeniz* cast off and began threading her way through the maze of crisscrossing Bosphorus ferries, out past the exotic Istanbul skyline punctuated with graceful mosques, spiky minarets and the elaborate outline of Topkapi Palace. The waterway widened into the Sea of Marmara and narrowed again in the Dardanelles as darkness fell.

There were 10 English-speaking passengers (8 UK) on this voyage, and we shared a table with two entertaining couples from England. In addition, the mostly Turkish passenger list included Turks from Cyprus (North), Israelis, Germans, and a scattering of other European nationalities.

Dinner the first night proved to be typical of the fare throughout the voyage - lentil soup with cream, fillet of sea bass and potato, aubergine, ground lamb and tomato, and a honey pastry, all washed down with a light domestic red or white wine. Breakfast invariably provided juice, bread, jam, cheese, a hard-boiled egg and tea but no coffee. Bland and uninteresting, the food turned out to be the one disappointment of the entire trip. No matter, I had come for the ship and not an upmarket cruise experience. Happily, the air-conditioning worked here and in the lounge during what would be recorded as one of the hottest summers on record.

We spent a pleasant first morning on the vast upward sloping foredeck seated on wooden deck chairs and sun loungers arranged alongside two cargo holds.

Above: the author stands at the bow of the Akdeniz.
Below: a Turkish charter boat comes alongside the Akdeniz at Marmaris hoping for some business.

Above: the Akdeniz's bridge is a feast of navigational instruments, some dating from the mid-1950s. Below: the Akdeniz is docked at Alanya on Turkey's south coast. Opposite: even at this hour the full moon is shining on a port with temperatures above 100F (38C) degrees.

In the evening we had drinks here and listened to a well-known (to the Turks) singer who sang from the deck above.

Arriving at the port of Izmir, the officers appeared on deck and on the bridge wearing their uniform hats as the *Istanbul*, the regular coastal ship to Istanbul, sailed by while exchanging whistles.

After sailing from Marmaris, a port with ferry connections to the Greek island of Rhodes, I joined the chief purser in the main lounge, the only other air-conditioned room on the ship. He had spent 30 years with TML, and the *Akdeniz* was the only non-ferry passenger ship now (with 750 berths) and had not carried cargo for the last four years. She could no longer sail abroad because of her non-compliance with international standards. Her engines were good but there was no air-conditioning in cabins and few had private facilities. Third class had six- and eight-person cabins and two dormitories, so for the three classes there were three dining rooms and three bars. The radio officer characterised the ship as like an old steam train in many ways but not operated by steam, then he added rather proudly, in rough seas the ship acts like a fish.

However, she was still profitable, and the main draws were reasonable fares and good service, and the passengers came for the seaside, varied towns, ancient ruins, and the backdrop of mountain scenery. He indicated the passenger list was typical - Turks, Germans, Israelis, Cypriots (N), some Italians, French, English and few Americans, or in this case, I was it and Suellyn the sole Australian.

The other ships in the fleet were the long-distance ferries *Ankara* sailing along the north coast and *Samsun* on the international run from Izmir to Venice, carrying commercial vehicles and Turks living and working in Western Europe.

The purser finished up by telling me that best ship in the fleet had been the previous *Ankara*, American-built in 1927 and acquired by Turkey in 1948. She was apparently a beautifully riding ship. Swan Hellenic chartered her for years for their cultural cruise programme, and she lasted well into the 1970s.

I recall meeting Swans (past Swan Hellenic passengers) on the subsequent ship, the *Orpheus*, who loved to tout their heritage, especially those that had occupied dormitory space on the *Ankara*. Now I can do the same when meeting Swans who only know the *Minerva* and *Minerva II*.

Alanya, the turnaround point, was perhaps the most memorable port with a very active waterfront. The population explodes from 30,000 in winter to 250,000 in summer. The locals came down to the pier to peer

through the portholes into the dormitories and crew quarters. Some tried to get aboard and a few made it. A near full moon rose behind the town, a fitting setting for a city that had been given to Cleopatra as a wedding present.

Leaving port, there was a sense of utter stillness when standing at the great upward sloping bow and looking back at the bridge, deck gear and kingposts, all silhouetted against a moonlit sky. Then I felt the quiet throbbing of the engines and slight flexing of the bow as the lights of the port gradually receded.

The nights on the south coast were still, very hot and humid with the overnight temperature barely falling below 100F. At bedtime, there was a slight breeze but with the cabin very hot and the fans not helping much, I took showers at bedtime and again during the night, while some of the dorm passengers bedded down on the open deck.

There were several important ancient sites between here and Antalya - Side, directly on the coast with Perge and Aspendos just inland. Perge had a stadium, theatre and Hellenistic Gate with the oldest existing towers,

Top: the cannon atop the fort at Bodrum has the Akdeniz squarely in its sight.
Above: one berth of our lovely twin-berth panelled cabin on the Akdeniz. A second class outside cabin. Note the privacy curtain enclosing the lower berth.

an impressive colonnade, a drainage system and lovely mosaics, some discovered by scraping away the dirt with one's foot. The last named boasted the first Roman theatre in the world (2nd century AD); it was preserved and still in use, rich with ornamentation, royal boxes and an arcade at the top.

At the port of Fethiye, the English-speaking contingent hired a boat and sailed along a very rugged coastline, pocked-marked with caves, then stopped for a swim. We shared the protected bay with the Fred. Olsen's *Black Prince*, riding at anchor with her swimming platform extending from the stern.

Bodrum provided a castle dating from the Crusades, its image used for the ship's menu covers. It was floodlit at night towering above the small harbour packed with its famous namesake trading ships, gulets, now mostly used for chartered coastal cruising.

The call at Kusadasi provided access to Ephesus, the ruin of ruins in this part of the world and a site that one never tires of tramping through as there is invariably something newly restored to see.

By 15th August, 10 days after leaving Istanbul, we were back at the same berth and ready to spend some time in the city before re-joining the delightful *Stella Oceanis* to finish the cruise we had started two weeks before. The return route sailed via Kusadasi, Rhodes, Heraklion, Santorini, Delos and Mykonos.

After disembarking at Piraeus we rode the narrow gauge train to Patras, embarked in the very attractive overnight ferry *Fedra* to Ancona, connected to a train to Milan and sleeper to Paris. After a few days here, it was train to Boulogne, Hovercraft to Dover, train up to London and finally back from Southampton across the Atlantic in *Queen Elizabeth 2*. This would be the longest and most complex round-trip surface travel itinerary through Europe that I had ever made, bound for a ship I had wanted to sail aboard since I began leafing through Laurence Dunn's 1961 edition of 'Passenger Liners'.

The *Akdeniz* passed to the Istanbul Technical University as a cadet training ship, but after several years being laid up, went for scrap in Aliaga in 2015, the last pocket liner of a once outstandingly beautiful fleet.

Above left (top to bottom): dormitory accommodations, not much privacy and hanging laundry. Because of the heat, many slept up on deck. Evening light and cooler temperatures while at sea northbound to Istanbul. Breakfast in the first-class dining room and coolish fresh air before the ship heats up during the day. Above: arriving back in Istanbul after a ten-night round voyage to the south coast.

Chapter Seven

Ivaran's AMERICANA

Passengers and Containers Down South America Way

15th April - 11th May 1989

The MS *Americana* was quite unlike any other passenger-carrying ship sailing the seven seas. Most cargo vessels offered just 12 or fewer berths; any more than that and a doctor must be hired and added to the crew list. Such 12-passenger ships used to number in the hundreds, sailing under the American, British and many European flags. Chasing freight, unlike passenger liners, they offered more leisurely offbeat itineraries and multiple days in some ports.

The onslaught of containerisation, pre-packing the cargo in steel boxes, rendered the time consuming break-bulk dockside handling of most cargoes obsolete. As shipping lines increasingly fell under flags of convenience, manned by third-world crews to cut costs, most cargo lines stopped carrying passengers. While some container ships still do, the number carried might be as few as two on up to eight or maybe 10, a rather limited group with which to be stuck for several weeks.

Ivaran Lines, a venerable Norwegian-flag shipping line that had traded between the U.S. East Coast and Brazil, Uruguay and Argentina since 1925, designed a container-carrying and refrigerated cargo ship that could take 86 passengers. A high standard of accommodation was housed in a seven-deck stern 'castle', while up to 1,120 steel containers could be stacked on deck or lowered into deep holds.

A passenger certificate (50 and above) classified the *Americana* as a passenger vessel, granting her preference in securing docking space in notoriously congested South American ports, hence speeding up freight schedules.

Ivaran went to a South Korean shipbuilder, and if this first ship was successful, two similar units would follow, most likely to be named *Brasilia* and *Argentina*.

Completed at Hyundai Heavy Industries in South Korea, the 19,203-gt *Americana* measured 580 feet (176.8 metres) in length with a service speed of 18 knots. She joined Ivaran's 12-ship fleet, two of which, the 1978-built *Salvador* and *Santa Fe,* carried up to 12 passengers.

Delivered in 1988, she arrived in New York harbour to berth along the Brooklyn waterfront just south of

The Americana passes through New York's Upper Bay at the start of a long voyage south to the East Coast of South America in May 1989.

Above: a roomy double cabin with a curtain divider between the sleeping area and lounge.
Below: enjoying pre-dinner drinks on the afterdeck with the Danish doctor (centre), and passengers are disembarking from the ship via a steep gangway.

Brooklyn Heights, long a prime shipping location that was rapidly losing ground to the more spacious container facilities in Port Elizabeth and Port Newark.

It would become a major undertaking for a freight company to sell the greatly increased number of passenger berths, but the owners expected the new ship's unique design and high standard of accommodation to propel it to success.

Upon completion, invitations went out to the travel press, and the response was so strong, including from top publications usually very choosey about events they might attend, that two chartered coaches transported the scores of the curious to the Brooklyn docks.

Immediately hooked by the beautiful cabins, public rooms and deck spaces, I wasted no time in lining up multiple magazine and newspaper markets to then secure a cabin for a mid-April 1989 sailing northbound from Buenos Aires back to Brooklyn. The one-way voyage would take more than three weeks, a long time to be away, but I did not want to short change myself nor my readers by booking a short segment and missing out on the rhythmic cycle of days at sea.

The main restaurant at breakfast and two maritime scenes at the forward end.
Opposite: the view over the containers from the bridge, nearly always open for passenger visits.

I flew south to Buenos Aires via Miami and Rio, and at the Lancaster Hotel reception, I immediately recognised the Irish brogue spoken by the delightful man with whom I had booked my room over the phone. He was a redheaded, third generation Irish-Argentinean.

While exploring Buenos Aires on foot, I found the city to be utterly enchanting with its wild, eclectic architecture mixed in with traditional European styles. I snapped more photos than I ever had before of any city. B.A. was a throwback to an earlier era, with people on the streets more traditionally Old World in their dress than Europeans in Europe. Especially noted were women wearing real furs to the opera.

Argentina had peaked economically in the 1920s, but since then long periods of stagnation and currency woes curtailed renewal and development. Besides the architecture, the subway cars were positively ancient, and the once robust British-built railway network had been allowed to deteriorate to a handful of rail lines mostly serving upmarket suburbs.

Ivaran's passenger agent was a Danish-born man named Mogens Buhl, who for 36 years had been the passenger agent for Blue Star Line's now withdrawn fleet of refrigerated cargo ships that had each carried upwards of 50 passengers. During the B.A. turnaround, they spent a full week loading frozen beef for Britain. Learning that I was a ship enthusiast, he gave me a beautiful set of promotional ship photographs that had decorated his passenger office. Subsequently, he acted as passenger agent for Delta Lines' Santa M quartet of 125-passenger-cargo container ships, and when they were retired, he became Ivaran's passenger agent. His duties had dwindled considerably and with inflation, he said that he had slipped from middle class to working poor, necessitating selling his car and no longer being able to enjoy nights out on the town with his wife or travel aboard.

On 15th April, I mounted the *Americana's* gangway at the container pier not far from the hotel. The passenger accommodation was located aft, while ahead lay the principal revenue earners - containerised and refrigerated freight. Most of the cargo on the route originated in Buenos Aires and Brazilian ports, little came southbound from the U.S.

The cabins and public room interiors were courtesy of the design team that produced the earlier *Sea Goddess* pair, hence they were luxuriously fitted with enough stowage space for a diva making an around-the-world cruise. There were 32 outside double cabins, some with roomy seating areas separated by a drawn curtain. A few opened aft onto Spartan metal decks, surely a design flaw, resulting in a dark and dreary atmosphere and of little use.

Cabin amenities included a coffee table, sofa and chair, two built-in desks with chairs, five closets, generous enclosed shelf space, a refrigerator, mini-bar, TV and VCR, combination safe, hair dryer, fresh fruit, flowers and a welcome-aboard bottle of Champagne.

Every bathroom had a tub/shower combination and bidet, generous storage shelves in the mirrored medicine cabinets, and a lavish gift-kit of toiletries. It was a most promising beginning.

Anticipating an older clientele, the ship offered no less than 20 single cabins, eight outside and 12 inside. Surprisingly, the cheapest per person passage was a berth in an inside single, something virtually unknown in the cruise business where solo travellers occupying doubles usually paid from a minimum of 125% to 200% of the per person double fare.

Two owner's suites offered two bedrooms each, two baths, and a separate lounge with three windows, truly palatial for a passenger-cargo vessel, let alone most cruise ships at the time. With an upper berth in some cabins, the passenger capacity could actually reach 104, but never did, by a long shot.

The main lounge with ample conversational seating and sit-up bar occupied the public room deck's port side, and adjacent a good-size, well-stocked library provided comfortable reading chairs and tables for playing card and board games. Curiously, it was little used except to take out books and videos.

The restaurant, in a mirror image location on the starboard side, handled all of the passengers at a single open seating. Except for breakfast where passengers arrived over a window of time, lunch and dinner drew most when the bell first sounded. A bit of competition for the tables sited next to the large windows caused some angst, with low-volume muttering but no real action taken.

For dinner men were asked to wear a jacket and tie and women dresses unless we were docked in port, a typical routine then. As I usually travelled light with just one-carry-on bag, and did not mind washing clothes along the way, I did not bring many changes of clothing. So I donned my single Madras jacket and added a tie to whatever shirt I was wearing, often a solid colour polo shirt.

Typically, breakfast and lunch offered a generous buffet that appealed to North and South Americans, along with Norwegian specialties (for the officers) such as marinated herring, smoked fish and that country's cheeses. Menu items were also offered. Dinner provided a generous selection within each course from a largely continental menu - with Argentinean beef and Chilean seafood recurring as main courses. There was only praise for the food preparation and the service.

Outside, aft and high up, the Sun Deck had a swimming pool, whirlpool and plenty of open space for shuffleboard and lounging, and occasionally a grill-style barbecue presented an alternative lunch venue taken at tables set under umbrellas. A small gym had basic exercise equipment, and a narrow passage offered two slot machines, while on two nights, casino gambling was set up in the main lounge.

The Crossing of the Line ceremony was done with true Norwegian flair, not unlike crossing the Arctic Circle aboard the Hurtigruten ships. The most willing passenger, and victim, was a 92-year-old woman who lived not a mile from my father, and she became my favourite table companion along with two couples from Australia. Mealtimes were a delight sharing stories of long and happy lives.

The *Americana's* crew numbered 52, comprising Norwegian officers and South American deck and catering staff, mostly drawn from Argentina and Chile, with additionally a couple of Brazilians for their Portuguese. Carrying more than 12 passengers, a Scandinavian doctor and nurse were in attendance, and the purser/cruise director, an American, handled the entertainment (group games and local musicians), shore excursions and individual tour arrangements.

While he was a bit over the hill and had worked mostly Caribbean cruise ships, he tried his best to appeal to the much older passengers than he was used to. His wife, traveling as a passenger, got much too involved in his job and soon she took on the nickname of Lady Macbeth and played the role all too well.

During the first part of the northbound voyage, cargo-handling ports came one after another, and the occasional sea day was much appreciated.

Sailing from Buenos Aires with 36 passengers, the overnight passage across the mouth of the River Plate was characterised by much thumping under the hull (cavitation) as the shallow muddy channel, missing many of its buoys, was constantly silting up. Deep draft ships often had to bypass the port altogether.

The ship arrived at Rio Grande do Sul at an awkward hour for going ashore, quickly loaded mostly tobacco, then pushed on to Itajai, a major Brazilian river port. We sailed past the city centre to berth just upstream. While mostly shoes and sardines were hoisted

Opposite: relaxation high up on the Sun Deck. Above: a bollard provided a perch at the bow for a quiet interlude before dinner.

aboard in some 81 containers, a chartered bus took us on a day excursion to Blumenau, a city settled by German missionaries in the 19th century. They had arrived at the mouth of the Itajai River and sailed inland as far as they could before establishing a settlement.

Remarkably, the city has retained much of its German heritage, and for a couple of hours, we could have been in the old country with the many blond, light-skinned residents speaking their ancestral tongue in a setting of their homeland timbered architecture. It was all a bit surreal in tropical Brazil.

The following port, Santos, was by far the largest, necessitating two full days dropping empty containers and loading full ones, with some left behind because of work slowdowns. The port was accessed through a narrow entrance flanked by a fortified jungle-covered headland facing the modern city on the opposite bank.

It was an uphill drive to Sao Paulo through an ever expanding, encircling slum populated by those who had left the countryside to exchange rural poverty for its urban equivalent. The city had rapidly become one of the world's most populous and most polluted.

As a sharp contrast, the rich lived in houses surrounded by high fences and locked gates, with male escorts hired to take the children to and from school due to possible kidnapping. Curiously, the city had the largest colony of Japanese outside Japan, the result of substantial immigration in the early years of the 20th century.

Lunch atop the Italia Building, then the tallest at 46 stories, gave a sweeping view over a grimy-looking city. We then had a brief sightseeing drive and a return to the ship at Santos.

Sailing on up the Brazilian coast, Salvador (Bahia) provided an intensely sharp contrast with 80 percent of its population of African origin. Officially founded on 27th April 1500, it established itself as the Portuguese colony's capital city up to 1763.

The original centre, now a UNESCO World Heritage Site, was made up of colourfully crumbling colonial buildings. What a contrast to the almost exclusively European nature of Buenos Aires, and given a single day's visit, it was impossible to make a relaxed transition to such a completely different, yet highly intriguing culture.

The final call before putting to sea was Fortaleza, the city centre located north of the container port, and characterised by a long beach front and a lively market, housed in a former prison, displaying a superabundance of fish, fresh produce, and lovely craft items made of lace and embroidery. A few of us enjoyed an excellent fresh fish lunch facing the beach.

When the ship departed Fortaleza, we looked forward to a nine-day passage direct to New York, and the true rhythm of the sea voyage took charge, a blissful experience now virtually cut off from the rest of the world.

I found myself spending a lot of time in shady deck locations reading and communing with the sea. Day after day our little world was surrounded by a horizon of puffy clouds and an occasional dark vertical tower indicating a squall. If it came our way, its intensity was usually brief. The stormy conditions might also create a vivid rainbow just before we scurried for cover.

My all-time favourite location was at the bow ahead of the containers. The trek to seclusion began by descending to the Main Deck, then opening a door to the outside ahead of the bridge and passenger accommodations. Walkways ran along both sides of the ship, sometimes passing under the stacked steel containers labelled with Ivaran, Hapag-Lloyd, Maersk, Mitsui-OSK, P&O-Nedlloyd and a dozen other shipping lines. Ahead of the containers, an open foredeck provided a seat on one of the flat bollards or on a ledge that flared out from the bulkhead.

This self-contained open space was out of sight of the bridge, and rarely did anyone interrupt the nearly silent isolation. If a light swell was running, one could feel a rising and falling motion with the ship flexing to the rhythm. I particularly enjoyed looking straight down at the bulbous bow, its knobby shape protruding from the hull.

One day a pod of dolphins suddenly appeared, clustering close to both sides of the bulbous bow. One rolled slightly over on his side with its right eye looking directly up at me, during which time the others shot off and abruptly returned. The encounter lasted about 15 minutes.

Other diversions were flying fish, quite a frequent sight below and above the Equator. I enjoy skipping flat stones at the seaside and thrilled in counting the number of times a silvery flying fish, much like a stone, used the crest of a small wave for staying airborne.

One evening, a full moon rose from the sea about an hour after sunset, and on the next it was too late as the dinner hour intervened. Later when I returned the moonlit setting was magnificent and no artificial light spoiled the starry sky.

Sailing on northward, the Southern Cross and Scorpio gradually slipped downward a bit more each night, and then the Plough (UK) or Big Dipper (US) appeared just above the horizon. Soon, the southern constellations vanished into the sea.

Before I retired, I would call in at the bridge as the officer of the watch was very welcoming and liked company. The helmsman was more taciturn and seldom spoke unless spoken to. I enjoyed learning about navigation, looking at the charts, having more constellations pointed out and hearing stories including lots about container theft and labour woes ashore.

The peace and quiet of the northbound voyage was shattered one night when an intense storm seemed to come from nowhere, while we were close to Bermuda.

Holding on tight, the ship corkscrewed for several hours, and the sudden lurching sent table items I did not get to soon enough to the floor. I decided against going above to see the action this time, the movements were so random and unpredictable. Happily, a few hours later we resumed a steady course.

Apart from the above sudden squall, as far as I was concerned, the cycle of days and nights could have gone on forever. The end came abruptly when the ship sailed into New York's Upper Bay docking at a Columbia Street pier on the Brooklyn waterfront.

My wife was standing on the pier, and after hearing tales from the deep, three years later in August 1992, we both made a similar northbound voyage, every bit as delightful as the first. The passengers were even more compatible, and we visited several couples for a number of years afterwards.

Sadly, the *Americana* never made it as a successful passenger carrier, averaging about 35 passengers a voyage and with very few South Americans booking. She made a profit as a cargo carrier, and the gravy was to be a healthy passenger load. As container traffic increased, the *Americana* could no longer handle the demand while some of Ivaran's other ships carried 50 per cent more. Originally the *Americana* was to have been joined by at least two more similar ships but they were never built.

In May 1998, CP Ships, formerly Canadian Pacific, bought Ivaran Lines, which had become unprofitable. CP's Ivaran name was replaced by that of Lykes Lines in 2000, and in late 2005, CP Ships was bought out by TUI AG and merged in mid-2006 into the Hapag-Lloyd organisation. Venerable shipping names vanished forever.

During the *Americana's* Ivaran ownership, the Brooklyn berth was replaced by one at Port Elizabeth, and when the ship was too small to fit into the route's rotation she was transferred to New Orleans undertaking a Caribbean circuit as a feeder ship. However, she was not successful in this role and made her last passenger sailing from New Orleans on 21st May 1999. Chartered for various trades as a container carrier with the lovely passenger accommodations sealed up, she carried the name *Golden Trade* when owned by a South Korean firm and flew the Bahamian flag. She was chartered to Safmarine for a while then became a shabby-looking tramp. In January 2010, she was sold to Chinese breakers.

It was a sad end to a failed experiment, one that would be likely to deter anyone else from trying to revive the once much loved combination passenger-cargo ship.

Opposite: shuffleboard on the Sun Deck as four Aussies look on, and lunchtime on the Sun Deck.
Above: the Americana is docked at Buenos Aires in August 1992 before sailing north along the Brazilian coast and onto New York.

Chapter Eight

RMS ST. HELENA

To the Remote South Atlantic

13th July - 16th August 1993

Islands make wonderful destinations, even more so when you can arrive by sea. Way back in March 1968, I visited the Seychelles when the only way to get there was by monthly ship from Bombay or Mombasa. The stay on this Somerset Maugham-style tropical island, effectively cut off from the real world, made me long for a similar experience.

Eventually I found it - St. Helena, a British dependent territory located in the South Atlantic some 1,200 miles off the coast of Angola. The mountainous 47-square-mile island qualified as one of the world's most significant places with no airport access. For its roughly 5,000 inhabitants, the 6,767-ton Royal Mail Ship *St. Helena* is their only link to the outside world, operating the longest regular liner service on the high seas.

Purpose-built in Aberdeen, Scotland in 1990 to serve all the island's requirements, the 'RMS' as she is affectionately known, then sailed from Cardiff, South Wales offering the facilities of a small liner for up to 128 passengers and with a British and St. Helenian crew of 65. The new ship replaced an earlier and smaller 'RMS' that faithfully plied the 6,427 nautical miles from Britain via Tenerife, Ascension to St. Helena and then onto Cape Town from 1978 to 1990. For decades prior to that, the mighty Union-Castle Line held down the route until the company bowed out of the passenger business in 1977.

With the island threatened by abandonment, a Cornish shipping entrepreneur, Andrew Bell, came to the island's rescue with a proposal for a passenger, mail and cargo service for a negotiated fee on behalf of the British Government. Now after 15 years, the popular service was well established.

In early July, my wife, Suellyn, and I boarded British Rail's high-speed InterCity 125 at London's Paddington Station for the 145-mile, two-hour journey to Cardiff. From the railway station we took a short cab ride to the docks.

The 'RMS' had been having engine problems and our sailing was some two weeks late, so before sailing south, the ship underwent speed trials in the Bristol Channel to test a new set of turbo blowers

Heading ashore by tender after a two-week voyage into the South Atlantic, (photo: Bruce Heard).

designed to bring the pair of Mirrless diesels up to full power.

Eight hours later, shore-side engineers disembarked via the pilot boat, and Captain Martin Smith announced that we were on our way with a scheduled arrival at St. Helena two weeks from today.

The 345-foot 'RMS' has four decks of passenger accommodation situated aft, while up to 1,500 tons of cargo are carried forward, variously in containers, as breakbulk and refrigerated.

The cabins, most with private facilities, are of average size, outside with windows or portholes and have two, three or four berths. Additional budget accommodations for 28, mostly inside cabins with wash basin only, are designed primarily for St. Helenians at subsidised rates.

Ours was a typical twin-bedded room with a window on A deck, furnished with a chest of drawers between the beds, two wardrobes with hanging space and shelves, a vanity table-cum-desk, and a chair. A

Above: the author is reading 'The Last Outposts of the British Empire' while sailing to one of the last aboard one of the last. The officers are engaged in blindfolded frog racing with 'jockeys' whispering advice. The frogs belong to the purser, a former Union-Castle man.
Below: the forward lounge aboard the RMS St. Helena is empty during the afternoon quiet period.

The dining room aboard the RMS St. Helena. Note the ocean liner prints on the far wall, (photo: Bruce Heard).

two-channel radio provided music and hourly BBC news broadcasts. A framed lace portrait of the island's indigenous wire bird decorated the wall.

An attractive forward-facing lounge spanned the width of A Deck with seats for 66, a sit-up bar, non-smoking section and a small library corner with books, videos and games. On the Promenade Deck above and aft, the Sun Lounge had seats for 60, a bar and a buffet for continental breakfast and light lunch as an alternative to the main restaurant. Large windows faced the heated saltwater pool and the open deck's tables with umbrellas, chairs, and sun loungers.

The restaurant, located on the starboard side of C Deck, offered two seatings with the top officers hosting several tables at dinner. Ours was Geoff Shallcross, the chief purser, who had formerly worked for Union-Castle. He was a lively conversationalist and gave us a wonderful time. On most nights, a jacket and tie was requested for men and the majority complied.

The fare comprised good British cooking with additional Indian curries. The soups, pâtés, desserts and cheese board were excellent while the best entrees were pork roast with crackling, roast beef and Yorkshire pudding and the midday Sunday curry. Reasonably priced South African and French wines began at just over four pounds a bottle.

Framed prints of Union-Castle, P&O, Royal Mail, Cunard and White Star liners lent a sense of history and decorated the walls and stairway landings, while a collection of ship models were on display in the purser's bureau.

The 79 passengers on Voyage 15 South were a varied lot. About half were St. Helenians, known as 'Saints' for short, a people of mixed British, South Asian, East Indian and Madagascar origin. They included students studying at British universities, and others returning from jobs in the UK and the Falklands or resettling in their island home.

The British governor's son was aboard and so was the French Consul who looks after the property that Emperor Napoleon Bonaparte occupied during his final six years in exile. Among the contingent en route to South Africa were a Swiss couple migrating to their newly-acquired farm, honeymooners from Johannesburg and a score of tourists from Britain, Germany, France, America and Australia.

Stored below decks were 93 bags of ordinary mail, registered mail and parcels, refrigerated, frozen and tinned food, pig feed, medical equipment and drugs, planks of a West African wood resistant to white ants, educational textbooks, nine automobiles and one-fifth of a tonne of postal stamps. The last on the list provided one of the island's few sources of revenue. While the

Left: a Saint and his Rhodesian ridgeback take a walk on the foredeck en route to St. Helena. Above: an upper and lower berth outside cabin aboard the RMS St. Helena. Below: the baggage barge at Ascension Island, (latter two photos: Bruce Heard).

livestock pens were empty, the kennel held a Rhodesian ridgeback named Ruby belonging to a father and son who had recently purchased one of the island's three hotels.

During the cycle of gradually warming days, the daytime routine began with morning tea at 7:30am, followed by either a light breakfast in the Sun Lounge or a full English breakfast in the restaurant, including, on occasion, kippers and black pudding, the latter definitely an acquired taste.

Morning activities involved reading and socialising on deck, visiting the bridge and taking a swim. After lunch, the ship went into a quiet period followed by afternoon tea with crust-less sandwiches, pastries and biscuits. Most fun was when a contingent of the more energetic officers and passengers gathered on the forward cargo hold for a fast game of deck tennis played with a hard rope or soft rubber rope ring, the former particularly tough on one's fingers.

Other daytime activities included clay pigeon shooting over the stern, pool games, tugs of war, deck quoits tournaments, a rousing crossing-the-line ceremony and travel videos geared to our destinations.

After dinner, the purser had the uncanny knack of getting everyone involved in one of his creative activities. For an icebreaker, he brought out his frogs - wooden cut outs tethered to a line that 'jockeys' gingerly coaxed across the floor during a series of spirited betting races. The trick was to find the right rhythm, for anything more than a gentle tug toppled the frog on its back.

The various races were made up with women or men only, one each from different nationalities, those with odd professions and lastly the officers. Perhaps the most fun to watch was all participants blindfolded to then be guided by a coach. This last race lasted as long as all the others combined.

One night we divided into teams of six for a trivia quiz that began with one member throwing a dart to determine the value of each question. Then as the weather warmed we enjoyed an evening of skittles out on deck using bowling pins and rope balls. An American-Australian couple from New York lost in the finals by a single point to two young Royal Navy officers.

During the six-hour bunkering call at Tenerife, we walked to the market and had lunch off the ship. About 20 passengers disembarked, mostly British living here or starting holidays. Five new faces appeared, a couple on contract to evaluate St. Helena's school curriculum, a French TV team on assignment to produce a documentary on Napoleon's last years in exile, and a German friend of mine who simply appeared without any warning.

After Tenerife, the 'RMS' left the main shipping lanes for the remote South Atlantic, and ship sightings dropped to an occasional tanker or container vessel. The sea remained placid, erupting occasionally when flying fish and porpoises broke the surface. We passed within a mile of the western tip of Senegal and even closer to a Black Star Line freighter wrecked on a reef. Then as dusk fell, the lights of Dakar brightened into a glowing backdrop for a deck barbecue and dancing. Later a violent electrical storm raged over the land, providing a dramatic spectacle viewed at a safe distance.

Opposite: deck activities included shooting, tug of war, deck tennis and, of course, the Crossing the Line ceremony.
This page: Suellyn Scull is reading the ship's programme seated in an outside twin-berth cabin. The ship's small pool, (photo: Bruce Heard).

Twelve days after leaving Cardiff, Ascension came into view, a desolate and forbidding volcanic island rising 2,800 feet above sea level from the base of the mid-Atlantic ridge. The ship anchored overnight to take on 40 passengers, a few of the hundreds of Saints who work under contract at British and American bases here and in the Falkland Islands. The British expanded its military presence after the 1982 Falklands War with Argentina and that required hiring additional civilian workers.

While 20 of us planned to go ashore with the purser, the sea began to kick up and heavy swells made it unwise to negotiate the ship's ladder, a pontoon landing and the slippery stone steps ashore. In addition, the captain wanted to sail as soon as possible, given that the ship was a fortnight late.

Two days later at dawn, the island of St. Helena spread across the horizon beneath low clouds clinging to the mountain peaks. Brown at the edges and green in the centre, the island's steep cliffs provide a natural fortress. The 'RMS' dropped anchor off Jamestown, a pastel-coloured eighteenth and nineteenth-century Georgian town sandwiched into a deep valley that slices inland to the island's incredibly beautiful central highlands.

The Saints had arrived home, and the through passengers got the bonus of a week on the island while the 'RMS' discharged its cargo and made a four-night passenger run to Ascension and back. Landing at the slippery stone steps was tricky business, and the less agile climbed into a wooden box, which was crane-loaded into a barge and then hoisted up onto the pier.

Gail Thorpe, who would provide us with accommodation, met us, and we walked through the original East India Company gate embossed with a wire bird, the island's only endemic species and related to a plover. We would soon see them out on Deadwood Plain where the airport will eventually be constructed, destroying part of its habitat, one of several factors that have resulted in the wire bird being placed on the list of endangered species.

Most visitors stayed at two small hotels in Jamestown, while we, and a few others, chose bed and breakfasts or self-catered cottages up in the hills. We hired a car for the week and hesitatingly negotiated the twisting one-lane tracks and blind hairpin curves that led to our farm and a three-room stone cottage, built from a former cowshed. It sat in a setting of sloping pastures, fields of shimmering flax, bordered by wild arum lilies and thickly wooded valleys. The countryside was Scotland, but steeper and with a better climate. The roads were such that only once during our stay do I recall getting the car into top gear.

A rising swell prevented the 'RMS' from discharging more than a fraction of its cargo into lighters, so the ship sailed early on the passenger shuttle run to Ascension.

During the week's stay on the island, we visited Longwood where Napoleon spent his last six years. He designed the gardens with sunken paths to prevent the British guards from watching his every move on his daily strolls. When he finally died in 1821, he was buried on the island until 1840 when his remains were taken to Les Invalides in Paris. Longwood House and grounds are French territory, overseen by the French Consul who arrived on our sailing.

Top: the timid swing ashore in The Box, and visitors come to our former cowshed.

Above: a classic view of the RMS St. Helena anchored off Jamestown before embarking passengers for Cape Town on the following day.
Below: Jonathan is by far the oldest resident on the island (162 years old in 1993), seen here with admirers on the grounds of Plantation House, the governor's residence built in 1791.

The British governor's wife, a Saint, invited the transit passengers for a tour of Plantation House, a handsome 1791-built mansion set in lovely grounds. One of its inhabitants is Jonathan, a tortoise who arrived on the island from the Seychelles in 1882 at the age of 50, making him 162 years old in 1993. He just may be the oldest reptile on earth.

No day passed without a trip into Jamestown with its attractive commercial and residential main street and the island's social centre. We ordered bread one day and picked it up fresh the next morning along with what fresh food we could buy. The list ran to green beans, spinach, carrots, tuna, wahoo and island beef. Our hosts, the Thorpes, generously provided fresh vegetables from their garden. The food supplies from the 'RMS' were very slow getting into the stores.

We spent three days hiking the island's many miles of walking paths along the coast and down to secluded bays, enjoying picnics, and sharing the day with ship friends who were staying in hotels. The island has virtually no sandy beaches but its beauty and isolation more than make up for that.

The radio broadcasts kept us abreast of overseas news, social events and when the 'RMS' would return to finish the cargo handling and embark passengers for Cape Town. We would have liked to stay longer, and were sad to leave the island, our hosts and our stone cottage tucked away up in the green hills.

Leaving St. Helena for Cape Town, the small complement of in transit passengers and a few newly embarked Saints enjoyed five more days at sea. It gave us time to share stories of our week on the island and to prepare for a visit to South Africa.

Refrigerated containers occupied our deck tennis court so one of the group got the idea of creating a smaller version over the after stress hold, using piled up furniture for a net, and the daily matches continued. We saw the green flash one evening, and the expected rough weather in the Roaring Forties never materialised.

On the last evening, no one wanted to leave the deck, because going inside indicated the end of the voyage. Just short of four weeks after departing Cardiff, the little RMS *St. Helena* sailed into Cape Town's Table Bay as dawn broke over Table Mountain. The ship would remain docked for three days to load fresh food, supplies, cement and coal, then return northward.

We spent a wonderfully varied fortnight in South Africa visiting Cape Town, Cape Peninsula, the wine country, riding the Blue Train and Rovos Rail and enjoying game drives at Kruger National Park and at Ulusaba, a private reserve.

The St. Helena airport was completed in the spring of 2016, and the first scheduled commercial flight was due to land on 21st May. However, problems with wind shear have temporarily scuttled those plans and the ship has been granted a temporary reprieve while the issue is resolved. When that happens, the ship will be sold and the final true Royal Mail ship service will come to an end, no longer linking by sea one of the last remote places with a significant population of just over 4,000 to the outside world.

Footnote: as this book went to press, the 'RMS' had scheduled sailings from Cape Town to St. Helena and Ascension through to July 2017 due to wind shear problems at the newly constructed airport preventing the start of weekly air service.

Opposite: charabanc for touring, the RMS St. Helena's funnel marking was often referred to as the South Atlantic prawn, and a life ring aboard the RMS St. Helena, (latter two photos: Bruce Heard).
This page: Saints going off-island and finally the RMS is docked in Cape Town in the shadow of Lion's Head.

Chapter Nine

SS ROTTERDAM

Last Atlantic Crossing

19th - 27th April 1997

I began a long association with Holland America's SS *Rotterdam* in 1964 when I joined the company in the Fifth Avenue passenger office. After less than a year, I moved to the headquarters on Pier 40 to be a part of the newly formed Sales Promotion Division. Our ships tied up right outside my window, so I had ample opportunity to scamper around them.

I moved on in 1966 when I boarded the SS *Rotterdam* on an eastbound crossing to begin a year reading for a Master's Degree in history at the University of London. The voyage, shared with about 800 graduate students, sealed my love for the ship, her rich interior décor that did not even hint at being derived, and her ingenious layout that transformed her with ease from a two-class ship in transatlantic service to one class for cruising.

In April 1995 I made the Atlantic crossing from Lisbon to New York, and two years later I booked what would be the last one she would make under Holland America as she had been sold to Premier Cruises to soon become the *Rembrandt*.

The powers to be at HAL wanted a largely uniform fleet, and the one-off *Rotterdam* no longer fitted their new formula; choose any 'Dam' ship, they are all the same. If an intending passenger, not knowing the ship, wanted to book the *Rotterdam*, perhaps simply because of an attractive itinerary, the agent would have a lot of explaining to do - she may be our classic ship but she does not have this and that and she is old.

On 17th April, I flew to Lisbon and spent the night in an inconveniently sited hotel where Holland America had booked group space. Then on the 19th, I boarded the ship at the Alcantara Terminal just south of the city centre.

In a nearby wet dock, *Princess Danae* was being refitted, a ship noted for her lovely hull lines derived from her days as the Port Line freighter *Port Melbourne*. The motor sailing vessel *Wind Star* sailed at 5pm and the venerable *Sea Cloud* arrived an hour later, the latter

The Rotterdam is berthed at Ponta Delgada, Azores on during a transatlantic crossing to New York.

showing lots of rust streaks and corrosion. Both ships had just crossed the Atlantic to begin a Mediterranean season after wintering in the Caribbean.

I occupied Cabin 258, a large, twin-port-holed outside on Lower Promenade Deck positioned well aft, sharing with a long-time friend from New York who lived just two blocks from Pier 40. Two beds were placed far apart with a curtain to draw at the head end for privacy at night. The space was fitted with a table, two chairs and stools, double dressers and cabinets, and three closets, as there was a folded away third upper berth above mine. Wood trim outlined the bulkheads, and pretty batik bedspreads, pillowslips and curtains displayed lots of colour. The Indonesian patterns reflected the large contingent of Indonesian crew (plus Filipinos) that had replaced most of the Dutch years ago. En suite facilities included a stall shower and heated towel rack.

The ship remained in port until the next day, and that morning a 95-year-old passenger suddenly collapsed and died in the Ocean Bar. His body was disembarked and we sailed at 12:05pm. The departure took us first down the Tagus, under the great suspension bridge and out to sea, westbound for New York with an island call in the Azores.

Top: the Rotterdam berthed at Lisbon on 7th April 1995. Above: the wide promenade deck and deck games forward of the funnels.

Top to bottom: the author on the Rotterdam's bridge during the Atlantic crossing, a double cabin with a curtained off sitting area and Captain Jakob Dijk in his quarters during the final Atlantic crossing for the Rotterdam.

I immediately went to the souvenir shop and found that a memorabilia dealer had largely stripped the stock of SS *Rotterdam* items and especially the commemorative end-of-era tiles. With many ship aficionados aboard, this was a great disappointment and did nothing to elevate the dealer's reputation when he began reselling the items at higher prices.

A ship enthusiast aboard presented a slide show of the *Titanic* film set, a three-quarter-size model built along the Mexican coast. It exhibited an amazing attention to detail, especially notable in the grand staircase, woodwork, and cabin decks. The talk's finale showed the set being demolished. On the following day, a most intelligent talk entitled the 'Floating Palaces of Art Deco', discussed ship design and interior décor.

On 22nd April, the ship called at Ponta Delgada, the largest city in the Azorean island chain. A group of us hired a taxi for a lovely coastal drive that followed twisting, narrow roads lined with stately trees, high hedges, a profusion of flowers, and especially the brilliant colours of azaleas. We stopped in several villages, each anchored by a Catholic church and with streets lined with pretty white houses topped with red roofs. Upon sailing from the island, Captain Jakob Dijk warned of rough weather ahead but not much materialised.

One morning, a Holland America custom had the captain, chief engineer, and hotel manager gather for a panel discussion. One passenger, thinking he was clever, asked, 'Why captain, did you not wear a life jacket during the boat drill'? His answer, 'I have to stay with the ship; my job is to stay out of the water'.

Another passenger asked about the rogue wave incident some years before off Casablanca. His answer, 'The ship rolled 36 degrees, and there was 2.5 million dollars in damage'. The ship was sailing in a location where there is a sudden change in the depth of the sea, and large swells suddenly piled up and sent the ship into a steep roll. Lots of deck chairs went overboard.

As I was writing what is now known as a blog for Travel Weekly, an industry newspaper, I wanted to capture some of the passengers' remarks about the ship. Some, paraphrased, ran like this.

Passenger One offered: 'This ship is virtually unchanged, while on most older ships, very little is original as new owners try to modernise by wiping out the past. Aboard the *Rotterdam*, the original concept and atmosphere have remained pretty much the same; she is a time capsule and my ideal. This is what I was looking for'.

Passenger Two: 'I always wanted to do an Atlantic crossing, what this ship was designed for. I sailed before on a cruise but this is different, and most important, I met my wife here in 1981 when it was already my favourite ship'.

Passenger Three: 'I wanted to be on this ship. She is in her element, and I wanted to say goodbye. It will be hard to transfer my affection, but I guess it will be *Norway* because I sailed on her as the *France*'.

Passenger Four: 'Five or six ships have or had souls: *Mauretania, Ile de France, Queen Mary, Nieuw Amsterdam*, and this one. She is the last of what a luxury liner should be - beauty and quality that cannot be duplicated. She is the most successful passenger liner ever built with a record of safety, number of ports visited and passengers carried, and she has made 20 full round world cruises'.

On the 25th April, my parents' anniversary, I woke up or perhaps I should say was abruptly awakened by my cabinmate standing over me and quite firmly saying, 'Get up, this is what we came for'. It was 6am and immediately I was aware that the seas were running high.

Climbing to the Boat Deck, several familiar faces had already gathered at the forward end, port side and behind glass. The air was filled with spray with the sea rolling directly at us and then slightly quartering to the port side. Over the next few hours, the wind would increase to Force 9-10, about 45 knots, creating 35-foot waves.

At irregular intervals, an especially large wave loomed before us, and we tried to guess if it would result in white or green water over the bow, and if green, an intense shuddering accompanied the twin props rising near the surface. In succession, several waves slammed down on the bow with the thick spray peppering the forward windows and swirling along the cavity of the covered boat deck. When it was truly a big one, a great burst of cheering followed and then satisfaction smiles; it was enormously energising. We were a crazy little group, and being outside, and in the fresh air, we could see what was coming while those within could not.

Damage began to accumulate - burst pipes, a window shattering in the card room, water filtering from the decks down into the galley, the crew quarters, and even the engine room. Waves that came down hard on the foredeck tore loose several cables and other gear, so Captain Dijk rang 'slow' to maintain headway while the deck crew put things right.

Outside the shop, the deck was an absolute mess of broken liquor bottles, china, and fallen racks of clothes. The horizontal expansion joints caused the carpet to pucker and stretch, a strange sight indeed.

The Rotterdam plunges into a major storm during a last North Atlantic crossing from Lisbon to New York in April 1997, spray deflected by the bow gets sucked into and along the Rotterdam's Boat Deck, which slopes downward as the ship rises atop a swell. The contents of the shop have been strewn about following the major North Atlantic storm.

CHAPTER NINE

Not everybody understood that this was a good thing; some even thought the ship might be coming apart. Partition planters in the Lido fell over, and passengers were tossed onto the deck. At lunch, one of the few passengers seated in the restaurant flipped over backward, and audible sounds of crashing crockery became almost repetitious. There were reports of people injured, even hips broken.

On the whole, most passengers seemed generally quiet, waiting it out. Some loved it but others, who were not necessarily feeling ill, soon tired of it all during the course of the day. Many in the cabins were quite sick. I felt fine, except very briefly at lunch in the Lido. Happily, the sensation passed almost immediately when I stepped outside into the fresh air. All programmes were cancelled for the day, including Shirley Jones' appearance, the grand buffet, a lecture and films.

Some of the Filipinos, who were accommodated in the forepeak, were terrified by the sea slamming against the hull. They filtered up to the Boat Deck forward and made a sleeping nest at night using deck chairs.

During the night of 26th April, the weather moderated making way for a beautiful morning. Spirits were revived, and I heard few complaints other than the storm lasted too long, even from some diehard rough weather aficionados. For a small contingent, it was a great way to celebrate the last crossing. Now the voyage was drawing to an end and our estimated late arrival the following day was 1:00 pm with a Sandy Hook pilot boarding three hours earlier.

Top: lucky tourist class passengers enjoyed the Ocean Bar when the Rotterdam was formerly a two-class ship; now everyone can.
Above: the Library and the Odyssey Dining Room, which was formerly allocated to tourist class passengers.

Upon arrival, the City of New York's *Firefighter* gave us a celebratory squirt for the ship's last call, and soon the *Rotterdam* was waiting off the pier for longshoremen to start work promptly at 1 o'clock. Off the ship at 2:15, I took a taxi home on the Upper East Side, dropped my bags, tore through some of my mail, and by 3:30pm I headed back to the ship for a Steamship Historical Society of America event.

Before going aboard, I went up the pier's top deck and had a good look at the streaks of rust at the bow and along the vertical expansion joints. Yes, this ship had been through some heavy weather.

Top: the lovely Smoking Room (then non-smoking) depicting smokers in bas-relief from around the world on the far panelled wall.
Above: a detail from the Ritz-Carlton Room.

At the planned event aboard, Captain Dijk was presented with the S.S. *Rotterdam* Ship of the Year Award. He proudly announced that after the storm, the ship had made 25 knots, and after 38 years steaming around the world, he had no idea his ship could still do it. A spontaneous cheer went up, and the Captain Dijk's eyes filled with tears. He also shared that 45 of the 90 ice carvings crashed to the deck during the storm, and 60 were rebuilt or repaired after the storm. His next posting would be the new flagship *Rotterdam* (VI).

Our group filed off at 10:30pm and a small contingent stayed to watch the ship sail from New York for the last time. Farewell *SS Rotterdam.*

Postscript: The ship first operated for Premier Cruises until that line failed, and she is now serving as a hotel, restaurant, meeting place, office centre and attraction at the ship's homeport of *Rotterdam*. I have toured her, dined and stayed aboard, and she is well cared for in her static role.

Opposite: views of two sections of the Ambassador Lounge.
This page top: the slender uptakes of the Rotterdam and the Rotterdam before sailing at midnight from New York on 27th April 1997.

Chapter Ten

QUEEN ELIZABETH 2

Roughing it at Sea with Cunard

12th - 19th December 1998

Some people are terrified to go to down to the sea in ships, and a few that do never go back; I can understand that. I watched my brother experience days of agony then sink into almost unconsciousness on a particularly tempestuous Atlantic crossing aboard CGT's *Flandre* when we were both teenagers. He did not venture out to sea again, until forced to ten years later as a Marine Corps officer. Perhaps being in charge of 80 troops helped him sail through that hurricane off North Carolina, because in 1991 he fell ill again within hours of leaving port aboard the *Queen Elizabeth 2*.

That very early voyage decades ago ran into multiple storms, the first producing huge swells soon after leaving the Channel, two more in mid-Atlantic, and a full-blown hurricane named *Daisy*. She packed winds that reached 125 mph, crumpling the bow, shearing off railings, and smashing several promenade deck windows, notwithstanding all that was shattered and scattered within. I loved the experience, the novelty of it, not getting sick, having to feel my way along the sloping corridors, grasping lines stretched across open spaces, hardly seeing a soul about. I found a safe perch behind thick glass just below the bridge to watch the sea rise up before me to then slam down on the foredeck completely blotting out everything in green water turbulence. Still in secondary school, I had my essay about 'what I did last summer' nailed.

It was decades before I again experienced major storms at sea, other than on overnight packets in the Mediterranean and North of Scotland. The SS *Rotterdam*, 39 years later, responded to my wish in spades, on what would be her last Atlantic crossing under the Holland America banner. I covered that most memorable voyage in an earlier chapter.

But this story is really about Cunard and with my 51 Atlantic crossings, 30 of them in *Queen Elizabeth 2*, there was a good prospect of a second decent storm, given that a dozen of those crossings took place in December and January.

My first Cunarder was the RMS *Sylvania*, westbound from Liverpool on a millpond in June 1964. Three years later in October 1967, I came home after a year in London, via Southampton, aboard the fully booked RMS *Queen Elizabeth*, not long before she would

Above: the Queen Elizabeth 2 is outbound from New York on 25th June 2003. Opposite top: an inspection is taking place in New York on 24th September 1995 after the Queen Elizabeth 2 encountered a massive 96-foot wave on the westbound crossing depressing the forward deck. Bottom: Queens Grill cabins such as this were amongst the largest of the original accommodation aboard the Queen Elizabeth 2.

be retired. While it was rough enough to have waves explode at the bow and feel the spray when standing aft of the first funnel, it was the sounds that I recall most. The wind whistled through the guide wires and rigging, but far stronger reverberations emanated from within the liner.

The constant creaking was ear-splitting as the ship reacted to the punishing pressures. She slammed into

Above: the Queen Elizabeth 2 in New York on her 1,000th voyage on 14th June 1995. Opposite: the seas are running 35-40 feet high during a westbound December Atlantic crossing aboard the Queen Elizabeth 2 and passengers hang on the Boat Deck railing during a winter North Atlantic storm.

the swells and shuddered not unlike blunt-bow cruise ships do today, but the difference was that we were still steaming at 27-28 knots, and not just maintaining headway. The sudden lurching that ensued was as if standing on a moving plate, powerful enough to throw anyone into harm's way if not holding tight to something firmly rooted. The intense lurch and wobble repeated itself a half dozen times, accompanied by the wood panelling and internal structure creaking loudly enough to drown out all other sounds. What pressure the ship must have endured over all those years, to keep to a tight schedule week after week.

But my most frequent launch pad into stormy seas was the now (2016) sadly languishing *Queen Elizabeth 2*. To clear the deck, I was not aboard for the famous 96-foot wave that crashed down on the bow in September 1995, but I did head down to the West Side pier to see her come in and have a look at the indented bow decking and the location of the missing fog horn and tripod. That wave received lots of media attention, and the incident continued to get dredged up for years when reporters needed some past drama to add to more current 'storm of the century' purple prose.

The scene at sea that night, so vividly described by Captain Ronald Warwick, had him suddenly facing what he called the White Cliffs of Dover before the seas came crashing down onto his ship. That incident, now two decades ago, occurred late at night, and most passengers and crew had retired, so happily, I do not recall hearing about injuries. Inspected by the U.S. Coast Guard in New York, *Queen Elizabeth 2* routinely put to sea on her next crossing but what if the ship had been half the tonnage and without her inherent strength?

Moving forward to December 1998, my wife and I were ending a stay in Cornwall, visiting good friends whose elevated property facing southwest into the Atlantic might as well have been the bow of a ship. As we headed for Redruth station it was blowing a gale, and after Teignmouth where the line runs along the coast, the waves slamming against the railway embankment peppered the carriage windows with salt spray. It was just a foretaste of what was to come.

At Southampton we were soon established in a Queens Grill-grade cabin, one of the original midships staterooms that we loved. Back then, most lecturers received one of these in return for singing for their supper. Cabin 2071 was spacious with an elliptical window in three panes, generous areas of wood panelling, a cream-flecked purple carpet and white painted ceiling. The furnishings included twin beds pushed together, two chairs facing a small round table, writing desk and vanity

Above: passengers aboard the Queen Elizabeth 2 are out enjoying the storm. Opposite and page 90: the Queen Elizabeth 2 dives into an Atlantic swell during a tandem crossing with the brand-new Queen Victoria in January 2008, (these three photos: Jon Miller).

mirror, lots of cabinet and shelf space, walk-in closet in a separate space and a second closet in the foyer. The bathroom had a tub and it was possible to make a little circuit from the bedroom into the closet passage then through the bathroom and out to the foyer.

As we sailed, the weather was at first just blowy but then turned much nastier for the first half of the crossing, plunging into one low-pressure zone after another.

The stormy weather produced one of the silliest performances I have ever witnessed at sea. In the Queens Room one afternoon, some photographers had been hired to film the service of tea - a major event with Cunard - but picked a rather inauspicious time to do so. The prim and proper models of a certain age were seated at two tables with the staff poised to serve them. Problem: the slick marble table tops had no raised edges and as the staff began setting the tables, the ship went into a roll sending cups, saucers and tableware sliding off onto the floor. The carpet saved some, but not all, of the china from smashing. Then when it appeared that the rolling had subsided, the staff were instructed to reset. They began serving the tea and sandwiches, and the shuddering suddenly returned, this time sending the whole tea service onto the floor. The models did not attempt to rescue the toppling china; they just sat there and watched with the poor staff looking on in helpless horror. The shoot was then abandoned, and I sat there shaking my head in disbelief.

That night the seas became rougher, causing considerable pitching accompanied by heavy wobbles. I did not sleep well at all and, in the morning, the Lido buffet was closed and the shops had items strewn all over the floor. At first, the sea was running up from the south, but then shifted around to create rolling and pitching, before ultimately heavily pitching only. Posted signs warned passengers to stay off the outer decks. With no interior location to see forward other than via the cabin TV, we watched the waves explode at the bow with occasionally four or five very large ones in succession. The following night, I went aft on One Deck. The stars above looked down on a mighty frothy white sea. I kept a firm grasp of the railings as the deck was rising and falling fast, sometimes without much warning. Walking unaided across a wide-open space was out of the question. I

CUNARD

CUNARD

attempted that only once, quickly dropping down onto the deck when it suddenly went out from under me, and staying there until I could make for a railing.

After dinner, I went to One Deck aft, and as I turned around to look forward, a deck chair swept up by the wind was sent over the Sun Deck railing to land near where I was standing. It then slid all the way to the aft railing. After catching my breath, I pulled it to a sheltered spot and went inside to inform one of stewards to alert the bridge.

The next day started out foggy with the deep fog horn sounding its warning with rigid regularity. I sat for a bit in the Queens Grill Bar on Boat Deck, and then decided to take my horseshoe constitutional, exiting by a starboard door and down some steps to the deck. Walking aft the winds seemed peculiar; instead of continuing to push me from behind, the blast abruptly reversed and hit me face on.

Making my way around to the port side, heavy sheets of spray were being sucked aboard with almost every oncoming wave, so I headed back to the starboard side. Approaching the doorway I had previously passed though, a huge ventilator had come crashing down from the space above and landed on its side. A friend, sitting just inside, said he was looking out the window and saw the ventilator just miss me. The howling winds had prevented me from hearing it, even as it hit the deck. My friend contacted the bridge and soon two officers came along to carry the ventilator away. Later looking at the spot, I saw an indented tear in the teak decking. While it remained for the rest of the time the ship was in service, the jagged hole was eventually treated with filler.

The pitching, shuddering and wobbling, caused by the rough seas, and whipped up by force 10-11 winds made the panelled cabin quite noisy as the ship lurched from side to side. At 8:05 the next morning I was standing in the shower bath when there was a bang that nearly knocked me out of the tub. Shortly after, there was a Priority One announcement - Crew Mess, Deck One, and then that the watertight doors would be closing on Decks 6, 7, and 8.

Clearly something major was going on, and with my wife suddenly awake and wondering what had happened, I quickly got dressed and went to find out. I decided the Queens Grill was the place to go, as the staff knew me there. Sitting down at our table I casually started a conversation with the assistant restaurant manager.

Apparently, the weather door embedded in the forward superstructure had given way after the ship took a dive into a particularly deep swell, burying the forward part of the ship's superstructure and putting incredible pressure on it. The sea rose to the bridge level some 90 feet up. The crew mess was just behind that weather door, and happily most of the stewards had left for work; luckily only one man was injured. The water entered the ship and wiped out the crew mess. The weather door then slammed up against the exit door, and water began pouring down the forward companionway and along alleyways where the crew accommodations were located.

Later, a crewmember told me that he opened his cabin door and found himself looking straight at the scene in the recent movie "Titanic" (1997) when the sea began flooding the ship. In this case, it was very quickly contained but still did plenty of seawater damage. The ship hove to for several hours while the opening was secured.

One of the most frequent announcements, of the few that Cunard permits, was 'Niagara, Niagara' - a

A cover from the daily programme featuring the drawings of Laurence Dunn.

code to indicate flooding, followed by the location, deck and perhaps the stairway letter. Throughout the day, we heard this code with some regularity.

After breakfast, my wife said to me, 'You are in a state of shock'. She was correct because I was in overdrive trying to take it all in. I went above to Deck Two forward, passing two officers, one of whom asked the other, 'Has the water reached the Midships Lobby'? The other answered, 'Not Yet'. Apparently the water had entered into the spaces between One and Two decks and was flowing aft.

A man came out of his cabin as I was passing and suggested I have a look inside. Water was flowing down the bulkhead. Here and there passengers had to be moved to dry cabins. One of my friends later told me that he was sound asleep in his well forward inside

Ted and Suellyn Scull aboard the Queen Elizabeth 2 westbound to New York on 11 October 2008, and Ocean Books ran a wonderful bookshop on the ship, selling all sorts of quality memorabilia in addition to books.

Some photographs from the Queen Elizabeth 2's final year. Top: the Princess Grill, the Midships Lobby, the fine model of the Mauretania and the ship's bell in a forward stair tower.

single. There was a terrific bang and he found himself awake, on the floor beside his bed. He had no idea what had happened.

There was little discussion that I recall hearing around the ship other than remarks about the turbulent weather. Often, particularly on a large ship, most passengers may be unaware of specific occurrences, and that's usually a good thing. I kept quiet about what I knew. By mid-morning the ship increased speed to 15 knots, then an hour later to 24, and by the time the seas had somewhat calmed, we were making 28-29 knots in the direction of New York.

That afternoon I entered the theatre to give a talk to the passengers, and the starboard forward seating section was roped off with water was dripping down from the ceiling. I asked the projectionist if I should carry on, and he indicated yes. The ship was still moving about so I held onto the podium, and just as I was about to speak to a smallish crowd of about 75, a pipe burst at the edge of the balcony dumping water onto several young people. They squealed and quickly exited, and a few minutes later came back and chose drier seats. Later in the day, 'Niagara, Niagara' identified the Yacht Club; a pipe had burst above the band playing for the cocktail hour.

That night, I collapsed into bed dead tired and, as we were about to switch off the TV, a newscast announced that President Bill Clinton had been impeached. Happily, we would be in New York in the morning to catch up with what was happening in the world beyond our 'safe port at sea in a storm'.

The stalwart *Queen Elizabeth 2* has probably experienced the toughest life of any passenger vessel in the world, sailing day after day at high speed in almost any weather, and at the time of this crossing, for 30 years. She would have almost another ten years to go.

With 74 Atlantic crossings by ships of a dozen lines, only about six were truly memorable for rough seas. Now with the giant *Queen Mary 2,* twice the size of her predecessor and such a steady ship, I doubt I will again experience such dramatic effects from major storms. Life at sea has been good so I accept that enough is enough. To wish for more, is not fair to the ship, its crew and the majority of passengers, including my dear brother.

Above: the author has just disembarked at New York from the Queen Elizabeth 2 for the last time on 16th October 2008, (photo: Suellyn Scull).

Top: the Queen Elizabeth2 sails for the last time from New York on 16th October 2008. Above: the Queen Elizabeth 2 and Queen Mary 2 rendezvous in New York's Upper Bay before the veteran Cunarder leads her replacement through the Narrows and out to sea at the start of her final crossing.

Chapter Eleven

NORWAY

9 / 11 Transatlantic Crossing

4th - 14th September 2001

For some unknown reason, I was overly anxious about the upcoming trip abroad. One factor was most certainly the considerable length of time away, 25 days, but that was of my own doing. At least, I had enough writing assignments to make the venture worthwhile.

Early in the afternoon on 5th September 2001, I taxied down to Pier 90 to board the SS *Norway* for a nine-night transatlantic crossing via Halifax and St. John's to Scotland where I would disembark, while the ship and most passengers continued south to Le Havre and Southampton. It was designed to be a celebratory voyage for ship enthusiasts who loved the *Norway*, *France* (her original name) and the French Line. Norwegian Cruise Line was not revealing what would happen to the ship after she arrived in Europe, but as she was approaching 40 years old, the end might be near.

I had been on hand on a cold morning in February 1962 when the brand-new SS *France* arrived in New York harbour on her maiden voyage. She exhibited the peerless profile of a thoroughbred transatlantic speedster, vaguely akin to the *Normandie*, yet with much unique to this ship, especially her winged funnels.

On that cold morning, New York's French population gathered at the French Line pier to see her angle into her berth, and the half-dozen or so French restaurants that clustered just inland around 8th and 9th Avenues freely offered patisseries, croissants, sandwiches and strong hot coffee. It was a national celebration of the much-venerated French Line.

My love for ships came with my first voyage aboard the SS *Liberté* and the second, coming home on C.G.T.'s much smaller SS *Flandre*. Subsequent short-sea C.G.T. voyages took place in the Mediterranean followed by two outstanding westbound crossings aboard the SS *France* in 1968 and 1973. The French Line was very much alive - the food and wine, service and the fun of it all, even when travelling in a minimum-rate tourist class berth.

Most of *Norway's* passengers had boarded in Miami, including many who lived in the Northeast, in order to spend the maximum amount of time aboard the ship, so here in New York, with relatively few embarkations, there was little delay getting aboard.

My shared cabin had been high-end tourist class aboard the *France*, an outside with two widely separated beds, a double chest of drawers in between, three closets because of a third upper berth, and tub bath. The atmosphere was plain but spacious. While this block of cabins was located directly under the Promenade Deck,

The Norway is anchored off St. John's Newfoundland on 9th September 2001.

Above: the Norway is berthed on the West Side of Manhattan at the traditional ocean liner piers. Below: a portion of a double outside cabin.

joggers were asked not to begin pounding the teak until after 9am, though that was not always the case.

At sailing time, I was summoned to the bridge to provide commentary during the passage down the Hudson. However, we went nowhere except up against the adjacent pier. With three tugs at the stern and one at the bow, the ship reversed a bit then grounded and moved against Pier 92 shattering windows, mangling the railing and knocking a large dent into the side. The ship then moved forward and reversed again but suddenly stopped. A barge and tow were coming down river with the current, and they had the right of way.

Finally, the *Norway* was properly positioned in midstream and began her eastward voyage. I watched as a heavily sweating docking pilot disembarked. The day was simply beautiful as I began my commentary. The fireboat *John J. Harvey* gave a farewell salute with arcs of red and blue plumes as we paused before the Statue of Liberty to be photographed from a hovering helicopter.

We then we passed under the Verrazano-Narrows Bridge, dropped the Sandy Hook pilot, and revved up to the ship's service speed, more than ten knots below what the *France* would have been making a quarter century before. Speed at sea was no longer very important in winning passengers.

I joined a table for nine in the Windward Restaurant, the former first-class Chambord aboard the *France*. While the room overall appeared much the same, many more tables had been added making it quite crowded and unusually noisy, obviating any thought of speaking to those seated across the table. We were mostly New Yorkers plus a Bermudian, Australian and New Zealander, all known to one another.

Since I had been aboard the ship as the *Norway*, the first time on her maiden arrival in New York in May 1980, a lot of original features had been further altered. Now the ship seemed a bit junky and cluttered. Little seemed left of the *France* within apart from the Windward Restaurant and its grande descente and the two-level Leeward Restaurant. The Club International, a sensational space lost much of its natural light when the aft windows were blocked up. The redecoration, however, was admirable as much of the *France's* interiors had been rather ho-hum.

The big test came when late one night I ascended to a favourite location on the highest deck just aft of the second funnel, my daily ritual aboard the *France*. In 1968 and 1973, the nearly 30-knot speed and the towering winged funnels spewing white exhaust to either side provided a tremendous sense of power and purpose beneath my feet. The strong winds rushing at me muffled the sounds of slapping squash balls in the court one deck below.

Now with the *Norway's* two added cabin decks, the funnels were submerged and far less impressive, and two new vertical pipes sent the exhaust skyward relegating the wings of the stacks to redundant appendages. We were almost drifting across the sea rather than plunging forward on a headlong mission to reach France in under five days.

I did go up again when I wanted to be alone. However, I kept a low profile about my disappointment as few aboard seemed to have known her in the glory days.

We called at Halifax and spent a beautiful sunny day on the harbour walk, riding the ferry to Dartmouth and back, strolling through the lovely Public Gardens and spending the remaining time at Pier 21 for its immigration museum and evocative reminiscences by new would-be Canadians. Many would still be alive as the depot did not officially close until 1971.

Back at sea on the eastward trek, we paused over the site of the *Titanic's* grave to pay respects, then headed off to Newfoundland.

Given that almost half the passengers were French, I put together a talk on the French Line, its origins, expansion as a successful passenger carrier from French ports to the U.S. and to French interests in the Caribbean and Mediterranean. I covered the most notable liners, shifted into cruising and described the unexpected end in 1974 when the SS *France* was hijacked by the crew, then abruptly withdrawn and laid up. With my personal experience, I could relate the special atmosphere aboard CGT ships.

Opposite: while the Norway's funnels look dramatic, two added passenger decks have submerged the bases, considerably diminishing their grandeur. This page: the Norway anchored off St. John's Newfoundland on 9th September 2001, and the Northern Princess, a Canadian coastal ship (built 1966) docked at St. John's, served on the Gulf of St. Lawrence route between St. Barbe, Newfoundland and Blanc Sablon, Quebec.

7
DECK
SLIPPERY
WHEN WET

Opposite: the Norway's boat deck provides a good constitutional on what was once the world's longest passenger ship at 1,035 feet (315 meters). Above: our table in the former first-class restaurant Chambord (now renamed Windward).

Afterwards, the offspring of the *France's* last captain came to the podium almost in tears as they said that I had recaptured what the French Line was all about. That made all the effort more than worthwhile.

On two nights, I ran ocean liner trivia quizzes and quite a few participated with nine tables of four competing. I chose both easy and hard questions so every table would get some points. Both evenings appeared to be successful.

We dropped anchor off St. John's, Newfoundland and had a sparkling day ashore, briefly in town, searching for lighthouses and visiting Cape Spear, Canada's easternmost port.

Then everything changed at lunch on 11th September. Seated with our usual group in the Windward Restaurant, the captain came on the tannoy to say that a plane had crashed into the centre of Manhattan. That's all he knew and he would come back when he learned more.

In less than 15 minutes he came on again, speaking in English, and many French passengers kept talking, so I stood up. I recall rising but I have no recollection of what I said. My friends told me that I shouted out 'silence s'il vous plait' and they did. I was already in shock.

The news unfolded that a jet had deliberately been flown into One World Trade Centre hitting the tower high up and engulfing it in flames, then 18 minutes later another slammed into Two World Trade Centre.

We quickly left the restaurant and went to one of our cabins and turned on CNN. That initiated a marathon of watching the terrorist attack playing over and over again, resulting in confusion about what happened and in what order. As the news progressed and the towers fell, the focus became frustratingly narrow with little news about any buildings around the WTC such as Trinity Church, to which we belong, St. Paul's Chapel where we were married and any number of other notable buildings. That information did not come until several days later as did the number of causalities. Soon, we learned about the Pentagon attack and the flight crashing onto a field in Pennsylvania.

I desperately wanted to be back in New York, not here, but at least I was amongst friends and that helped a lot. I went up to the North Cape Bar and cancelled my talk, and the few people that showed completely understood.

The ship allowed us to have free Internet, and I was able to connect to one friend who lived next to the WTC complex. He had run for his life when the first tower collapsed. After reaching the Battery, he was evacuated by boat to Bayonne but with no money and

Above: the enclosed promenades are ideal for a winter North Atlantic crossing.
Opposite: the Norway's maiden arrival in New York in May 1980 and a postcard of the ship from the time of my crossing.

no identification. Having ingested this and other news, I felt even more depressed and isolated.

On 14th September, I was one of only two who disembarked (as planned) at Greenock near the mouth of the Clyde. I took the train into Glasgow Central Station and found an Internet café that, for over two hours of connection, cost just one pound.

My Lower Manhattan friend was now living on a couch at another friend's apartment and had no idea when he would be able to get back home. Another friend living in the West Village filled me in on what was damaged and destroyed. Miraculously St. Paul's Chapel right across Church Street from the WTC was spared even a broken window.

As my wife would be waiting for me in Passau, Germany, and there was no way to contact her, I carried on with my itinerary and that the same afternoon left Glasgow by train for Newcastle, connecting to an overnight ferry to Ijmuiden near Amsterdam and continuing south by train through Germany.

A friend living in Essen, who knew my itinerary, boarded the train at Duisburg and handed me a care package of water, apple, banana, and sweets. I barely held back from bursting into tears. He disembarked at Dusseldorf ten minutes later, and I changed at Cologne for a train down the Rhine and inland to Wurzburg, pausing for the night. I researched my city profile in the morning with a private guide then boarded a train to Nuremburg and onto Passau.

An Austrian friend who I had first met over 20 years ago aboard P&O's *Oriana*, greeted me at the station, having come three hours by train from her town of Wels. Together we walked to the riverboat *Mozart* where my wife Suellyn was anxiously waiting. We heard her story of being told about the terrorist attack while visiting a school in England, and at least she was with acquaintances.

As there were no planes flying home at this point we continued on our planned programme of sailing down the Danube. Happily, we shared a table with the only other Americans aboard as they too had been in Europe when the attack occurred.

The SS *Norway* had a somewhat troubled and shortened life when a boiler exploded in May 2003 killing eight engine crewmembers and injuring another 17. She was immediately taken out of service, and after many delays and much red tape she was eventually towed to the scrap yard in India. Demolition began in December 2007, marking the last and final connection to the French Line.

Chapter Twelve

Norwegian Coastal's LOFOTEN

Hurtigruten's time machine

April 2013

The Norwegian Coastal Voyage, known as the Hurtigruten, translating as Fast Route, is my all-time favourite ship trip. The venerable passenger and cargo service has operated for a century and a quarter, with a sailing from Bergen every night of the year for the 2,500-mile coastal journey well beyond the Arctic Circle, calling at 34 ports in each direction over a 12-day cycle.

With the rugged shoreline characterised by numerous fjords flanked by steep mountainsides and off-shore islands, uninterrupted land travel over long distances was next to impossible; hence a regular sea service was established in 1893 to link the South with the North of Norway.

Today, the daily year-round schedule requires eleven ships, and passengers may travel between any two ports and create pleasure trips by booking the full one-way passage between Bergen and Kirkenes, a small mining town near the Russian and Finnish borders, or simply stay aboard for the complete round voyage.

Over many decades the ships have markedly changed from tiny coastal steamers measuring just over 1,000 tons, to the much larger vessels in use today, in the 11,000- to 16,000-ton range. Simplicity has given way to relative luxury by providing comfortable outside cabins with en suite facilities, multiple lounges and meeting spaces for conferences. In addition, on ten of the eleven ships, the cargo rolls on and rolls off through openings in the ship's hull.

The eleventh, the diminutive 2,621-ton *Lofoten*, is altogether different, a present-day throwback to an earlier era and consequently much sought after by classic ship aficionados. On the other hand, it is best avoided by those who could not imagine sailing in a tiny cabin with upper and lower berths or, perish the thought, possibly scurrying down the corridor for a shower if the small selection of cabins with twin beds and en suite are all booked up.

With the *Lofoten*, the cargo is stacked on wooden pallets and then moved by forklift to the ship's side where a crane operator swings the load aboard and lowers it

Above: the Lofoten leaves Kirkenes, the turnaround port, for the one-week voyage south to Bergen. Opposite: double cabin 239 has an upper and lower berth at right angles and en-suite facilities.

into depths of the forward hold. What the ship transports is there for all to see - foodstuffs, household appliances, furniture, construction materials, tree saplings for replanting, in short, anything and everything.

In mid-April, I booked my sixth voyage up the coast some 47 years after first discovering this route. Then, with only a short vacation period available, I sailed in the 1949-built *Erling Jarl* from Bergen to Bodø, returning south by train. I was so enchanted by the ship's working function and cosy on-board atmosphere that I returned 20 years later to sail in the 1960-built *Harald Jarl* for the complete round trip. Then I was hooked, and for working assignments, I booked the *Kong Harald*, *Nordkapp* and a first sailing in the *Lofoten*.

Most recently I specifically chose the *Lofoten*, named after a Norwegian island chain, built locally in 1964, and measuring about a one-sixth the size of the newest vessels *Midnatsol* and *Trollfjord*. The *Lofoten* now represents the last of her type in regular service found anywhere in Europe or North America.

On this voyage, I was travelling with a friend, as my wife could not come. We first flew to Oslo and stayed two nights at the Thon Hotel, reasonably priced for Norway and within walking distance of Central Station when travelling with rolling suitcases.

With an Oslo Card, we visited Akershus Castle, its grounds and the WWII Resistance Museum, Nobel Peace Prize Museum, City Hall and its murals, Munch Museum, and Vigeland Sculpture Park using trams for longer stretches. We walked along Aker Brygge (waterfront) to find restaurants, of which there are many, many choices at various prices, all more costly than at home, but then this is Norway.

On a Sunday morning, we boarded the train for Bergen, booking Komfort Class, a small step up in fare with roomy seating. As it was April, snow appeared just an hour into the six-and-a-half-hour ride. By the time we reached the mountains, it was a winter wonderland, with blowing snow when we briefly stepped off for fresh air.

The café car was festive with many passengers heading to the ski slopes or visits to family and friends. Nearly everyone speaks some English and many are fluent so conversations ensued when we were recognised as foreigners.

Once over the pass, just beyond Myrdal where a connection is made to the steep railway line down to Flam, the snow turned to rain when the North Sea's coastal effect took hold. Arriving in Bergen just before 3pm, we walked to the centrally-located Thon Bristol Hotel in about 15 minutes, dropped off our luggage

Meeting at Molde, top: The Vesteralen is the second oldest ship in the Hurtigruten fleet, dating from 1983. Bottom: passengers from the Lofoten take a run ashore during the short call. Opposite: the bear skin displayed in the restaurant foyer, (photo: William Mayes).

and headed down to the waterfront to inspect the day's Hurtigruten ship, the *Kong Harald*. She was one of the first of the new generation (1993) of roll-on, roll-off ships. Visitors are welcome to board any of the company's ships when there are regular passengers aboard. Just show up at the gangway and secure a pass.

We liked the décor, enlivened by a collection of Norwegian paintings and sculpture, and took note of the larger cabins, wraparound panorama lounge, adjacent bar and long gallery lounge leading to the large aft-positioned restaurant.

On the second Bergen day, we explored the colourful, wooden Hanseatic League port buildings and rode the funicular up Mount Fløyen for great views over the city and out across the islands to the distant North Sea. We walked down along a well-marked path through the hillside residential area and on to the fish and produce market, partly housed in a modern pavilion at the head of the harbour. In the summer, the outside food stalls are a delight but on this chilly day we were happy to eat inside.

Boarding the *Lofoten* at Bergen the next mid-afternoon, we entered through the purser's reception with its well-worn polished wooden floors and panelled walls. A wooden staircase with tubular brass railings leads up to the Saloon Deck where a foyer lounge spans the width of the superstructure. A large white polar bear skin graces the wall between the restaurant's twin entrances.

Full one-way passengers (Bergen-Kirkenes) and round voyagers have meals included and dine in the main restaurant, a beautifully panelled room. Blue upholstered, wood-framed chairs are placed at tables of four and six, half set next to a window. Paintings of previous Hurtigruten ships decorate the walls.

Normally, apart from at the height of the season, passengers dine at one sitting, with breakfast and lunch served buffet-style, while dinner is a set three-course meal (no choice) at assigned tables. Special diets are accommodated with advance notice. We were placed at the captain's table, not that we ever ate with him. It was the private table for the master and his officers only at lunch. Our companions were two delightful couples, one from America and the other Anglo-American and living in Britain.

The continental menu reflects the tastes of the mostly European passengers but local smoked salmon, marinated herring and Brunost, a delicious sweet brown cheese, were available at breakfast and lunch. Grilled fish from the North Sea is naturally a popular entrée. The food is varied and well prepared, and I enjoyed the set dinners starting with soup or salad with main courses of grilled chicken or fish, roast beef or pork, and a dessert often accompanied by fresh fruit.

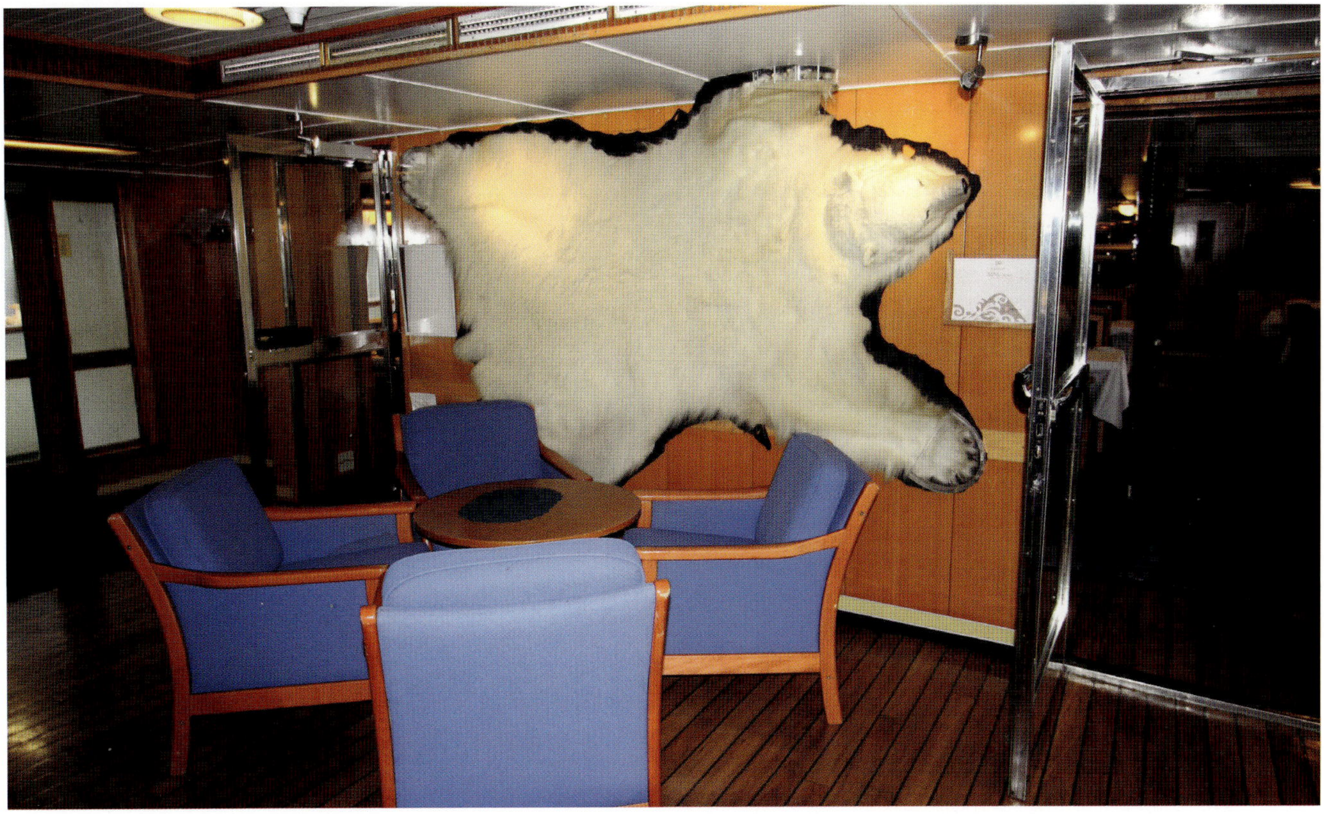

Interport passengers pay extra for meals, or frequent the adjacent cafeteria, a linear space running down the port side and connecting to an aft lounge bar. Both rooms are panelled and decorated with Norwegian ship and seascape paintings. Additionally, the cafeteria's gift shop sells Hurtigruten souvenirs.

We came prepared for very high alcohol prices; government-taxed as a luxury item. Instead of ordering pricey wine for dinner, I chose a pint of Norwegian beer for about $12. Some passengers bring aboard their own alcohol, and we had a nightly glass of wine in our cabin, sharing the tiny space with a friend who joined us from England and another from Switzerland.

Two forward facing lounges, spanning the width of the Boat and Saloon decks, provide ample upholstered chairs and couches for watching the

Top: our table taken by the husband of the woman to the left, (photo: a fellow passenger). Above: Norwegian herring and cheeses are accompanied by a wide variety of sauces and toppings at the lunchtime buffet. One of the many paintings of earlier Norwegian coastal ships is the previous Lofoten (1932-1964) that as the Kypros became a total loss after a fire off Haifa in 1966. Opposite top: a view of the almost empty restaurant shows the fine wood panelling, (photo: William Mayes). Bottom: the cafeteria, for passengers making short passages, also acts as a secondary lounge.

cargo handling and the sea ahead, playing board and card games, reading and socialising. The upper Panorama Lounge has the better view while the lower lounge (same deck as the restaurant) looks directly onto the forward deck with clear views to either side.

The open bridge wings on the highest of five decks are popular for watching the ship's progress as is the Boat Deck and a sheltered area facing aft (where smoking is permitted).

Cabin accommodation is the biggest hurdle to face, as the best cabins sell out early. Very few cabins have twin lower beds, and most are designed like a slightly enlarged railway sleeping compartment with upper and lower berths. On deck plans, categories N (3 cabins), J (3), A (20) and I (7) have private shower and toilet. The

Above: the Captain and First Officer on the bridge of the Lofoten.
Opposite: the staircase is another period piece, and a quiet afternoon in the aft lounge, (photos: William Mayes).

Ds have showers and toilets along the passageways. Total cabin berths number 154.

We had an A-category with an upper and lower, and with careful unpacking, we managed (just) with storage space in narrow closets, a few drawers and space beneath the lower bunk. Don't expect phones, TVs, radios or mini-fridges, the cabins are strictly for sleeping. It reminded me a bit of a college dorm room shared with a very messy roommate.

After leaving Bergen in the evening, the first important port of call, the next morning, is Ålesund, a city that experienced a horrendous fire in 1904 and which was completely rebuilt into today's handsome Jugendstil (Art Nouveau) style.

In the off-season, the port call lasts four hours allowing time to explore the town and make the 30-minute climb to a viewpoint overlooking the city and surroundings. Between mid-April and mid-September the ship pauses just 45 minutes before adding the four-hour passage deep into the spectacular Geirangerfjord, sailing between high cliffs and cascading waterfalls until the ship can go no further. The ship then returns to Ålesund, arriving in the late afternoon.

We booked the Geirangerfjord excursion up to the town at the top of the rim with the classic cruise-ship view down to the turning basin below. The coach, with the most inane commentary from our guide, continued over a scenic route back to Ålesund where, to her dismay, we left the bus at the viewpoint overlooking the city and walked back to the ship on our own, definitely the highlight of this outing.

Highlights from the commentary went like this. 'There's a typical Norwegian marina' (two boats at a tiny dock). 'It is busier in the summer. Look, here's a beautiful church to visit inside' (as we speed by without even slowing down). 'That hill ahead is to be blasted away for an IKEA shop, and look at the water standing on the ground because the ground is frozen.' As the bus approached a ferry, 'Once there were 80 factories and now almost zero, so people can work for the buses and ferries'. Passing an old farmhouse, 'That's 400 years old and now the owners are adding an inside toilet and shower. Look, Dagmar destroyed that fish farm but there weren't many fish because the people took them'. Scribbling as she spoke, I am not making this up. By the way, Dagmar was a cyclone.

Twice a day, an announcement heralds the meeting of the northbound and southbound ships, allowing about five minutes to appear on deck and not to miss the passing. Whistles are exchanged and photos taken in between waving to the southbound ship.

On the third day, the Trondheim call allowed nearly four hours (six in the off-season) ashore, so there is time to visit the southbound Hurtigruten ship, in this case the 1996-built *Nordkapp*, another contrast in size and ship design. We then walked into the city centre to Nidaros Cathedral (1030-1300), forever under repair, and returned past the pastel-coloured wooden storehouses lining the Nidelva River.

During the Arctic Circle crossing (Day 4), King Neptune came aboard to initiate the neophytes with a ladle or two of icy water poured down the back of the neck, followed by a restorative swig of cloudberry wine. A new stewardess was deluged with the entire

icy contents of a full bucket after which she went shrieking back to her cabin. These Vikings retain their brutal ways.

The ceremony was certainly an icebreaker, and by now we had met many of the other 70 or so passengers who hailed from Germany, France, Britain, Australia, Canada, U.S., and of course, Norway. The crew is mostly Norwegian and very friendly as they enjoy serving people who come specifically for this ship.

That afternoon included a two-and-a-half hour call at Bodø, an important fishing port and the start of the crossing to the Lofoten Islands, which appear like a formidable wall in the far distance. A number of deck passengers boarded for Stamsund, Svolvær and Stokmarknes, the last-named being Hurtigruten's headquarters and the location for the *Finnmarken* (1956), a preserved example of a traditional coastal ship. On the northbound the call is too late for a visit.

Opposite: a corner of the upper forward lounge with door leading to the central foyer. This page, top: while the Lofoten loads cargo, some passengers fan out to visit the town or turn around to photograph the ship. Above: a baggage rack for point to point passengers without cabins and the open deck, so close to fast-moving water, is a magical spot, (lower photos: William Mayes).

Tromsø made for another lovely stop with wooden houses and storefronts recalling the traditional Norwegian town appearance before fires, German occupation and resulting destruction during WWII. The four-hour call permitted a visit to dramatic A-frame-style Lutheran Cathedral accessed via a longish walk across a bridge, and also by excursion or local bus.

With the shoreline increasingly rugged and covered by more snow, Day 5 brought the excursion to the North Cape from the port of Honningsvåg. Standing on top of the promontory, I looked down and across the Barents Sea in the direction of the North Pole. Later in the day the ship rounded the North Cape for the final overnight passage to Kirkenes, the turnaround port.

Many passengers leave here for onward travel while round trippers return via the same ports but calling at different hours, so those previously scheduled at night become daylight stops. We rented a car for the day and drove to the Russian border, just fifteen minutes away and then into Finland, a 45-minute drive. Leaving the car at the airport, we headed south to Oslo on an evening flight. It won't be necessary to spend any more time in Kirkenes.

If sailing aboard one of the last examples of sea travel from half a century ago intrigues you, the *Lofoten* is your ship. Her charms are seductive, and be certain you know what cabin comforts you can do without. Book as early as possible to get the category you want.

Lofoten celebrated her 50th birthday in February 2014, and her many followers hope that she has many more years of service ahead. Hurtigruten's owners have decided to promote her as a nostalgic trip into the past, so it is likely that she will become better known beyond her present fiercely loyal aficionados.

Opposite: the Lofoten loads cargo the old-fashioned way with forklifts, wooden pallets and cranes, and a new crewmember gets the traditional initiation by King Neptune as the ship crosses the Arctic Circle. Above: in the far north of Norway, the southbound Nordlys slides by with an exchange of whistles. The Lofoten is docked at Honnigsvag while many passengers take the trip to the North Cape, 20th April 2013.

Chapter Thirteen

QUEEN MARY 2

Last Liner Linking Southampton and New York

1st - 7th May 2004

As a long-time transatlantic liner aficionado, I jumped for joy when Carnival Corporation purchased the venerable Cunard Line, almost certainly saving it from going under. The new owner wasted no time in announcing a project to construct the largest, tallest, widest, and most expensive passenger vessel in history. The new ship, to be named *Queen Mary 2*, would become the first true transatlantic liner since the *Queen Elizabeth 2* entered into service way back in 1969.

Naysayers thought that Cunard was out of its tree back then, more than a decade into the jet age, but the *QE2* had more than proven her worth and by the end of 2004 passed the *Aquitania* (1914-1949) as the longest serving Cunard liner on record, and she would go on to serve for another four years. The *QE2* captured my heart in June 1969 during a first crossing and remained a sea travel option for the entire time she was in service.

Happily, Carnival Corporation, the world's largest cruise company, had the financial resources to construct a one-of-a kind ship to undertake express Atlantic crossings for half the year and to compete in the pleasure cruise market at other times. Chantiers de l'Atlantique at St. Nazaire, France won the order and the contract was signed in November 2000. I visited the yard in December 2002 to check on the progress.

The new liner made her maiden voyage, from Southampton, in January 2004. I was a journalist travelling as a paying passenger on that first sailing, following a naming ceremony that could not have been more thrilling and performed by Her Majesty Queen Elizabeth II.

Several unique design specifications make the *Queen Mary 2* different from a warm-water Caribbean resort ship. A long, slim hull allows her to maintain an exacting schedule between New York and Southampton, England, to average just over 25 knots when called for, regardless of most adverse weather conditions. If she should fall behind, her considerable reserve power can drive her up to nearly 30 knots. In heavy weather, which we experienced one day, the liner's slender bow sliced through the waves, while in the same conditions, the blunt bow of a cruise ship would result in such pressure and stress to the hull that speed would have to be greatly reduced. For added strength, the steel hull from the keel to the Boat Deck is 50 per cent thicker, and extra care has been taken in welding procedures and stiffening the main bulkheads.

But there are concessions to increase her appeal as a modern cruise ship such as eight decks of balcony

Above: the 228-foot/69.5 meter clearance under the Verrazano-Narrows Bridge determined the height of the Queen Mary 2's funnel.
Opposite: the Britannia Restaurant may just be the grandest dining room at sea today.

cabins. Lest one conjure up a giant wave sweeping the partitions away, the three lowest balcony decks are recessed into the hull well above the waterline; they immediately got named 'hull holes'. In addition, a breakwater on the foredeck diminishes the impact of a rogue wave slamming against the superstructure, and I do not believe that it has been put to a complete test yet.

Having crossed the Atlantic by sea dozens of times since my teenage years, I had high hopes for the new ship, so following the maiden cruise voyage from England to Florida via the Caribbean, I then returned to make the second westbound transatlantic crossing from Southampton, England to New York in May of that inaugural year.

Arriving at the dockside terminal in Southampton on an early afternoon, I gazed up at the most imposing passenger ship one could ever imagine, at her towering yet graceful forward superstructure, her long slender bow and unmistakable Cunard red funnel. Then going aboard, initial impressions revealed public rooms, corridors and foyers on the same generous scale I recalled from long-ago voyages on the French liner *Liberté* and first *Queen Elizabeth*.

An open promenade, sheltered by lifeboats, stretches almost the entire length of the 1,132-foot/345-meter ship. With the deck laid in teak, handsome green-cushioned wooden steamer chairs line the railings or can be moved into cosy recesses between multiple bay windows. On warmer days, promenaders come out in force, some to chat with seated friends, an age-old social tradition at sea. My wife and I chose about 5pm or so to do our rounds; then we pretty much had the deck to ourselves.

Within, long interior sightlines extend the full length of the high and wide central passage, the walls framed by artful bas-reliefs depicting the four seasons on Deck 3 and images of the four elements on Deck 2. The former passes through the central atrium to a large portrait of the line's founder, Sir Samuel Cunard. On close examination, the ingenious effect is created by 7,000 miniature picture postcard depictions of the Cunard Line fleet through the decades.

Above: the Queen Mary 2's Boat Deck has one of the roomiest promenades of any liner or cruise ship, and three circuits make up just over a mile. Opposite: preparations for afternoon tea viewed from the G32 night club and afternoon tea in full swing in the Queens Room, (photos: William Mayes).

Standing at the entrance to the main Britannia Restaurant, a lighted dome arcs over a three-deck-high space flanked by three tiers of balconies connected to the main level below via a sweeping double staircase. Yes, you can make that grand entrance, but make sure you enter the restaurant via Deck 3 to be able descend to your audience.

A vertical tapestry by a Dutch artist, a richly colourful centrepiece, depicts a giant liner set against the New York skyline and Brooklyn Bridge. Unlike the ship's Grill Rooms, the main dining room has two sittings, yet I did not feel like second-class passenger given this grand setting. We joined friends at a second sitting table for eight on the highest tier next to a large window overlooking the sea; a good choice in the summer when there is still some daylight.

Dinner is reserved seating while breakfast and lunch are open. Our steward and his assistant were both from Poland and very accommodating to our requests. Two nights were formal, and nearly ninety per cent of the passengers complied, which is now not the case on most ships. But the truly grand *Queen Mary 2* deserves dressing up for. The other nights, other than the first and last, are informal, meaning a jacket and tie for men.

Equally impressive, the Queens Room, elegant, columned and two decks high, recalls a retro Art Deco ocean liner style with large windows looking out to sea. We liked the white glove afternoon tea service in the good company of a harpist, piano player or trio. Later in the day, the room becomes a party venue, a

Opposite: the well-stocked library, and the Queen Mary's Chart Room is one of the most evocative public spaces on any ship afloat. Above: the Queen Mary 2 in Las Palmas in January 2004, and the side galleries flanking Illuminations provide ideal spots to read while watching the sea. The photographer is about to occupy the one empty seat.

ballroom for serious dancing and to enjoy cabaret acts or Cunard's version of Ascot, complete with a fancy hat contest. Maureen Ryan, the ship's hostess who worked on the old *Queen Mary*, makes it a delight for participants and spectators alike. She is a genius at making fun of a passenger's costume without being the slightest bit mean.

The Chart Room, another oversize space and our favourite, is softly elegant with roomy seating, a sit up bar, large ocean-view windows and a wonderful cool jazz band playing during pre-dinner cocktails. Next door the Veuve Clicquot Bar, intimate and open to the atrium, provides a before-dinner glass of Champagne to accompany an order of caviar, paté de foie gras or smoked salmon.

The understated atrium rises through five decks, two of which house the reception and tour desks, and H. Stern, Hermes, and Dunhill; atrium-view cabins look out from the remaining three.

Terrific views over the bow can be had from the large, comfy library, the delightful, dark-panelled Commodore Club piano bar, the Atlantic (card) Room, an open observation deck below the bridge and from the glass enclosed Lookout on the highest deck. No ship afloat has so many wonderful forward facing viewpoints.

We are great readers, so the library is a daily destination to peruse some of the 8,000 books. If the weather outside is iffy or worse, take to an armchair for the bow view ahead or one of a small cluster of seats in the starboard wing, just below the bridge wing, with views in three compass directions. The library shop, as on the *Queen Elizabeth 2*, sells maritime books and ocean liner memorabilia. I am always on the look out for new Cunard Line lapel pins to add to my substantial collection.

In addition to the main Britannia Restaurant, the higher category cabins and suites are allocated to the Princess Grill or Queens Grill, two single-level rooms and mirror images of each other on Deck 7 aft. A Queens Grill Lounge and piano bar is reserved for the exclusive use of these passengers; otherwise there are

Above: the Queen Mary 2's kennel attendant is certainly in the right job. Opposite: the view of exploding seas from a comfy seat in Queen Mary's Library on Deck 8 and the author's brother Sandy watches the seas rising up before his eyes on Deck 2 forward.

no restrictions where one may go, apart from a small aft deck below the pet kennels.

Todd English, the big-name speciality alternative restaurant, features a creative Mediterranean-style menu designed by the Boston restaurateur. An early reservation is highly recommended, and in warm weather, lunch is served on the outdoor aft-facing terrace. Charges for lunch are $20 per person and dinner is $30; both we found well worth the cost.

For informal dining, the multi-sectioned Kings Court has several serving stations, with minimal queuing except at peak hours, for breakfast and lunch, offering some breakfast variations for American, British and European tastes. With a little searching, we were always able to secure one of the many bay window tables facing the side promenades; delightfully private seating away from the main traffic flow.

At night the Kings Court gets ingeniously transformed, using subdued lighting and temporary panels to screen the buffet tables, into four dining sections. Reservations are required but there is no extra charge in three of the quartet. The Carvery features succulent American cut roast beef; La Piazza, a typical Italian menu with self-service antipasto; and Lotus, our favourite, a highly successful Asian sampler menu. We sampled all three but not the Chef's Galley ($35 per person) where up to 26 participants watch a cooking demonstration then partake of the prepared meal.

On Deck 2, we enjoyed the Golden Lion with traditional English pub grub menu including fish and chips, bangers and mash, steak and mushroom pie. Spirited trivia quizzes take place here twice a day. Lastly, on a nice day, the Boardwalk Café on Deck 12 serves a grill menu with a score of tables set out on the open, and often windswept deck. This high-up open space becomes an outdoor cinema on a warm night to crunch on popcorn and watch a movie under the stars while ensconced in a padded deck chair.

Daytime activities feature a successful lecture and classroom-style seminar program put together by Oxford University on subjects such as history, politics, fashion and watercolour painting. We attended a series of four talks by a BBC radio commentator who talked about the early days of radio and how it handled breaking news and another two about the best American and British travel writers.

There are two theatres. Illuminations, attractively recalling a smallish Radio City Music Hall, becomes a planetarium later in the day offering three different shows. The plush red bi-level Royal Court Theatre hosts students from the Royal Academy of Dramatic Arts,

some big-name entertainment, cabaret acts and dance teams.

Deck space is far more generous than aboard the *QE2*, making the *QM2* a much better warm weather cruise ship. There are five pools, including one for children and one under a Magrodome. Children's facilities are excellent, and the baby-sitting received high marks for professionalism.

The smallest cabins (categories C1-C4 and D1-D5) are uniformly 194 square feet (18 sq. meters), with light

Above: the author is set up in the library to sell the first two volumes of ocean liner voyages. Opposite: the author is about to launch into a lecture in Illuminations beginning with sunrise at the Verrazano-Narrows Bridge, and the Queen Mary 2 approaches the Verrazano Narrows Bridge outbound.

wood-grained panelling and furnishings, adequate storage space, a small sitting area with pop-up table, chair and a vanity unit cum desk. Bathrooms have a shower, good counter space and limited shelves. An interactive TV shows films, documentaries, CNN, BBC and local programming when available, and is a repository of information about ship services, checking one's bill, tour information, menus, maintenance requests and sending e-mails using an infrared keyboard. Charges for outgoing (composing off-line) and incoming e-mails are a flat $1.50, much cheaper than typing a letter with the clock ticking at the Internet centres.

The standard outside cabins - the category we occupied - are somewhat larger than the interior portion of the cabins with balconies enclosed in the hull. Some found these latter to be dark, and the outer solid bulkhead prevents views of the sea when seated. Above the lifeboats, the balcony cabins (B1-B7, 248 sq.ft/23 sq.m) are more typical with Plexiglas instead of steel on the outer side, and larger interior sitting areas. Cabin sizes then increase to junior suites (P1-P2, 381 sq.ft/35.4 sq.m) with the largest (Q1-Q6, 506 to 2,249 sq.ft/209 sq.m) including aft facing duplexes. Those on Deck 8 have access to a concierge lounge.

The Canyon Ranch Spa, the largest such facility at sea, will undoubtedly be a major booking draw and received very high marks for the professional services. My wife loved her one-hour massage, which allows one to use the indoor pool and saunas for the day at no extra charge. It is located on two decks and includes a Fitness Centre, Aqua Therapy and a Beauty and Skin Care Centre. Services include a long list of massages, exercise for weight loss, back care, educational workshops, thermal therapies, scrubs, and cocoon and beauty treatments. Appointments are taken, charges include gratuities, and my wife said that there was no pushing to buy products.

On the maiden voyage, the ship took a Force 8 storm like an ocean greyhound with no reduction in service speed, and on the last leg from St. Thomas to Fort Lauderdale, we routinely steamed at 27 knots, hitting 28k for a short spell, with little vibration, even in my low-down aft cabin. On the transatlantic crossing, the ocean was like a millpond, and we got the thrill of the early morning arrival in New York harbour, a sail up the Hudson past the skyline, and a docking on the West Side.

The *Queen Mary 2* is herself a destination, and she will operate best on crossings and when combined

with cruises. There are so many places to explore and restaurants to sample that a simple six-night crossing is not anywhere near long enough. There are also some quirky design features, where fore and aft, some decks suddenly drop down half a dozen steps or more or rise an equal number. And it's a mighty long walk, for example, from the Library at the bow to Todd English at the stern.

It took me two weeks to feel that I understood the ship, and I found her to be an eminently worthy successor to the *Queen Elizabeth 2*, and as different from her as the *QE2* was from the original *Queen Elizabeth*. The sea option to and from England has sallied forth in grand ocean liner style.

Addendum:

We have returned again and again, switching my role from a journalist to a speciality lecturer making westbound crossings a couple times of the year and speaking on the topics of ocean liner travel and New York City as a destination.

Since the maiden voyage, there have been numerous changes, so while this account was designed to chronicle the ship as new, expect some differences.

Listed here are just a few, with some major alterations occurring during the substantial spring 2016 refit in Germany such as additional cabins, including most welcome singles. Todd English was replaced by The Veranda, recalling the original *Queen Mary's* Verandah Grill. The Winter Garden is now the Carinthia Lounge, with café and drinks services throughout the day. The Kings Court, the huge buffet restaurant, has been completely redesigned and is now far more attractive and easier to navigate. The number of pet kennels was doubled to 24 as they have been consistently sold out.

Cunard now sources all the lecturers, and new in 2015 are New York Times speakers on some eastbound crossings. The speciality restaurants open in the evening now carry a charge of $15. One significant change is lengthening the six-night, five-day crossing by a full day and sometimes two. And the wonderful hostess Maureen Ryan has retired, but she reappears for special occasions and her lectures about the early days working on ships are a hoot.

While some cruise ships are now considerably larger than the *Queen Mary 2*, she does remain the largest ocean liner ever built; with the difference between a liner and cruise ship described at the outset.

QUEEN MARY 2

Opposite: from the deck of the Queen May 2 docked at Brooklyn, an early dawn view under a full moon west to the State of Liberty. Above: an arcing rainbow frames the Queen Mary 2 on a westbound crossing to New York in tandem with the Queen Elizabeth 2. Finally, a brilliant ocean liner sunset aboard the Queen Mary 2 on the last night at sea before New York.

Index of Ships